know where the terms "fire plug" or "cherry picker" originated? Or why dalmations are fire dogs? Ever heard about the fireboat that received the Naval Distinguished Service Medal? What about the fireboat that kept the entrance to Pearl Harbor open after the surprise attack in 1941?

Did you ever hear about the mutual aid requested into Boston in 1872 for a fire that destroyed 776 buildings? That request brought 45 engines, 52 hose wagons, three ladder trucks and 1,689 firefighters. While Chicago burned on October 7, 1871 (remember hearing about Mrs. O'Leary's cow?) and 300 people lost their lives, another fire burned 250 miles away — at least 800 and possibly as many as 1,200 people were killed in that conflagration.

These little teasers only whet the appetite for the vast amount of history you'll find inside this book. We hope you will have a greater appreciation for today's fire service after you have read what others had to endure.

The Editors
Firehouse® Magazine

FireFighting Lore

Strange But True Stories from Firefighting History

by W. Fred Conway

Foreword by Paul Ditzel

Library of Congress Cataloging in Publication Data
Conway, W. Fred

Firefighting Lore – Strange But True Stories
From Firefighting History
Library of Congress Catalog Number: 93-090730
ISBN 0-925165-14-X

FBH Publishers, P.O. Box 711, New Albany, IN 47151-0711
© W. Fred Conway 1993. Second Printing June 1994.
Third Printing August 1996.

Printed in the United States of America

Front Cover Art & Design: Ron Grunder
Typography and Layout: Pam Campbell-Jones
Back Cover Photography: Mike Tonegawa – The "Chief Director" is
from the collection of the J.B. Speed Art Museum, Louisville, KY.

FOREWORD

Considering the broad influences the fire service has had upon so many facets of the history of the United States, it is curious that so few historians have delved into the subject.

One of those few is Fred Conway, author of *Firefighting Lore*. Conway, an eminently successful businessman, former chief of a volunteer fire department and a highly-respected religious and civic leader, has long had a profound interest in firefighting. It is out of a labor of love that he not only publishes books on firefighting history, but has made it possible for others to see their firefighting books in print which, otherwise, might never see publication. He is, moreover, an author in his own right; most notably his history of *Chemical Fire Engines*, which is the seminal work of its genre.

Firefighting Lore is a compendium of vignettes culled from the thousands of stories which have evolved since our earliest days. Few of them are intended to be profound in nature, but all of them deserve preservation as part of our nation's heritage.

Hopefully, the intriguing samplers in *Firefighting Lore* will stimulate other historians as well as firefighters and others interested in the fire service to seek out definitive writings, especially on the Great Chicago Fire, The Iroquois Theatre Fire and the fascinating history and development of those little red fire alarm boxes which were at one time as commonplace in American communities of every size as telephone booths.

America's firefighting history is replete with material which this book can only begin to savor. How many people know, for example, that Paul Revere, a volunteer firefighter, might never have been able to make his legendary ride without his firefighting credentials which gave him free access to the streets at night? That Benjamin Franklin founded the first volunteer fire company in Philadelphia and was instrumental as a founder of the first fire insurance company in the United States? (It's still in business!) It was Franklin, moreover, who coined that oft told phrase, "an ounce of prevention is worth a pound of cure" as he became a pioneer in espousing fire prevention.

At least several presidents used their volunteer firefighting status (which was a badge of social, economic and civic honor) as a stepping stone to the nation's highest office. George Washington was an active volunteer firefighter and frequently visited volunteer fire companies when he could spare time from his duties as commander-in-chief of military forces during the Revolution.

In America's larger cities, membership in a volunteer fire company opened up many avenues for members. Those who fought fires side by side in wintry blizzards usually formed business and political associations. Volunteers (and at one time there were some 6500 of them in New York City alone) often voted as a bloc and could elect mayors and governors. The social activities of volunteers were the focus of banquets, balls and weekend parties, picnics and sporting activities.

Volunteers who fought fires together marched off to war together. New York City volunteer firemen formed their own regiments which fought—and died—valiantly during some of the most famous battles during the War Between the States. Philadelphia's famous regiment of all-volunteer firefighters served heroically in nearly every major battle of the Civil War and a statue on the Gettysburg battlefield forever salutes their patriotism.

Firefighters were responsible for art and music. The lithographs of Currier & Ives grew out of Nathaniel Currier's volunteer firefighting membership and he is shown in one of the famous lithographs pulling a hose reel. Innumerable were the musical plays, dances and songs glorifying the volunteer firefighter.

Much of this changed after the War Between the States when cities began to disband volunteer fire companies in favor of paid firemen. One thing that did not change, however, was the incredible ability of firemen to find solutions to problems. One of the more fascinating examples of this is the creation of telephone stickers for quickly calling whatever emergency service was required: fire, police or medical. These were the creation of the author of this book who modestly tells how it all happened in *"The Story Behind Emergency Telephone Stickers"*; the concluding vignette in the book.

Wherever a firefighting need arose, it was, likely as not, a fireman who developed it. Frederick Seagrave was a Wisconsin builder of ladders for orchards when he saw that his volunteer firemen needed better and stronger ladders. Seagrave ultimately became one of the best-known manufacturers of aerial ladders and other fire apparatus in the United States.

Nor can we overlook my late friend, Bob Quinn, Chicago Fire Commissioner, who told me he was returning from a funeral. While driving through a park he saw an elevated platform being used to trim trees.

FOREWORD

Although a "cherry-picker" had been used to fight a major fire in Canada, it was Quinn who outfitted a borrowed picker and mounted large caliber hose and a nozzle. The apparatus became known as a Snorkel and soon found favor in fire departments throughout the world for its capabilities of doing some things that conventional aerial ladders could not. This intriguing story is told in Conway's chapter, *"What? Fight Fires With A Cherry-Picker?"*

Conway's collection of short but true stories is certain to fascinate not only those who are interested in firefighting, but open up new areas of historical interest among the vast numbers of those who know little about their fire departments but nevertheless have the highest respect for those men and women who, at a moment's notice, will risk their lives to protect those of others while serving as our mainstay against the elemental destruction of property by fire.

Paul Ditzel
Woodland Hills, California

Paul Ditzel is the author of 15 books and over 600 articles for magazines, including the Readers Digest. *He is a former contributing editor to* Firehouse Magazine *and* Fire Engineering. *His best selling book,* Fire Engines, Firefighters, *was submitted for a Pulitzer Prize in American History.*

ACKNOWLEDGEMENTS

Many of the stories in this book, all of them true, are the result of fire historians, preceeding me, who have ferreted out little known facts which would likely have passed into oblivion without their diligent effort. It would be difficult to list them all and to properly ascribe credit for each story. But all of these researchers, most of whom have delved into firefighting history as a labor of love, have my sincere gratitude.

The man who is himself a fire service legend, and who is not only the dean of fire authors but also the dean of fire historians, Paul Ditzel, was kind enough to review the manuscript for historical accuracy and to provide background material for several of the stories.

Thanks also to Louisville fire historian David Winges for supplying the material for *The Reindeer Hose Company Outing* and *Where's The Fire?* Lisa Parrott of the J.B. Speed Art Museum in Louisville, Kentucky, was extremely cooperative in permitting photography of the "Chief Director" for use on the back cover as well as in the story, *Where's The Fire?* My article, *Yesterday's Fire Extinguishers,* originally appeared in the June 1988 issue of *Firehouse* magazine, and I appreciate their permission to reprint it here.

My sincerest thanks go to my editor, my wife Betty, who took time to go over the first draft of the manuscript and make suggested changes.

Many of today's firefighters are unaware of much of their firefighting heritage. Hopefully this book will provide some historical background, both informative and interesting, which will give them a deeper appreciation for the traditions of which they are now a part, and which they can share with others. The role of a firefighter is truly a noble calling.

W. Fred Conway

OTHER BOOKS
BY W. FRED CONWAY

Chemical Fire Engines

—

Discovering America's Fire Museums

—

Corydon - The Forgotten Battle of the Civil War

—

The Most Incredible
Prison Escape of the Civil War

—

Squire – The Incredible Adventures of
Daniel Boone's Kid Brother

—

Young Abe Lincoln –
His Teenage Years in Indiana

—

What the Bible says about Prehistoric Man

CONTENTS

The Fire Dispatcher
Who Played The Guitar
As Chicago Burned

You have heard about the Roman Emperor Nero who played his fiddle while Rome burned; but have you heard about William Brown, the Chicago fire dispatcher who strummed his guitar, serenading his ladyfriend, while Chicago burned?

The date was October 8, 1871, and there had been scarcely any rain in Chicago for 14 weeks. Firemen had been run ragged, and on the previous day they had fought a mill fire for 17 hours.

About 8:30 p.m. 35-year-old Catherine O'Leary, who lived with her husband and children in a cottage on Chicago's west side, carried a lantern to the small barn back of her cottage to milk her cow. Setting the lantern on the barn floor, she began her milking when, for reasons unknown, the cow let loose with its right hind foot, which connected with the lantern. The fire which resulted soon engulfed the barn in flames.

Thousands fleeing the conflagration clogged the city's bridges.

In the third floor fire dispatching center at the Courthouse a mile to the northwest, Fire Alarm Dispatcher William J. Brown was busy playing his guitar for his sister Sarah and her friend Martha Dailey. Sarah noticed a glow to the southeast, but Brown continued to play his guitar, assuming it was but a rekindle of the previous fire and of no importance.

Meanwhile, Mrs. O'Leary's neighbors had formed a bucket brigade to fight the fire, but no one had thought to alert the fire department. Finally William Lee ran past a nearby fire alarm box all the way to Box 296 three blocks away. It was locked. In those days all fire alarm boxes were kept locked to discourage false alarms. Lee got the key from the corner druggist, unlocked the box, and pulled the lever. It was now 9:05 p.m.

Fire Dispatcher Willam Brown kept on strumming his guitar. Somehow the alarm didn't get through. No lights flashed, no bells rang — only the melodious strains of the guitar filled the alarm room.

At 9:21 p.m., nearly an hour after the fire started, with the glow on Chicago's night sky getting ever brighter, a watchman screamed at Dispatcher Brown, "Strike Box 342!" Brown obeyed, but Box 342 was more than a mile south of the fire. Three engines, two hose wagons, and two hook and ladders turned out — all heading in the wrong direction.

Realizing he had given the wrong location, the watchman again screamed at guitar-strumming Brown to strike the correct box, but this time Brown refused. He felt it would only compound the confusion.

"Why, my God, man, it was a terror to the world!"

As the fire spread out of control, Chicagoans stuffed bags with personal belongings and fled in terror.

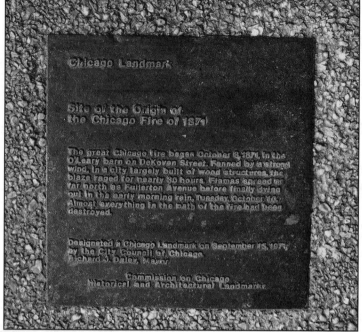

What could be more appropriate: Today, on the site of Mrs. O'Leary's barn, is the Quinn Fire Academy.

Someone again pulled Box 296. Obviously it was out of order — no alarm was received by Brown, who by this time had laid his guitar aside, realizing that he had a big problem. He took it upon himself to strike a second alarm, followed by a third alarm, and then a general alarm, calling out the entire Chicago Fire Department.

But it was too late. The comedy of errors led to the virtual destruction of the city of Chicago. Cities in eight different states sent engines, but to no avail. Mrs. O'Leary's lantern, the cow that kicked it over, the inoperative fire alarm box, guitar-strumming Fire Dispatcher Brown who ignored the glow, the watchman who gave the wrong location, and the 14 weeks without rain all combined to level much of the city of Chicago. Even fire companies arriving from Milwaukee, Cincinnati, Indianapolis, St. Louis, Pittsburgh, Philadelphia, and New York City could not check the flames before they had devoured an area five miles long and one mile wide, destroying over 17,000 homes and killing 300 people.

Each year in the United States, Fire Prevention Week is observed during the week in October which marks the anniversary of the Great Chicago Fire.

Which newspaper story was correct?

Was Kate O'Leary a seventy-year-old hag, or was she a thirty-five-year-old buxom Irish housewife?

Will The Real Mrs. O'Leary Please Stand Up!

All right, Kate O'Leary, which one are you? Are you a seventy-year-old hag or a thirty-five-year-old, tall, buxom Irish housewife? Chicago newspapers have described you both ways:

Tribune: The O'Leary's are the worthy old couple who owned the cow stable.

Times: Mrs. O'Leary was an old hag whose appearance indicated great poverty. She was about seventy years of age and bent almost double with the weight of many years of toil, trouble, and privation.

Journal: A stout Irish woman, some 35 years of age.

Times: [in a second article] She is a tall, stout Irish woman with no intelligence. During her testimony, the infant she held kicked its bare legs around and drew nourishment from mammoth reservoirs.

Journal: When the reporter suggested that the fire must have been pretty rough on her, she exclaimed, "Rough! Why, My God, man, it was a terror to the world!"

WHILE CHICAGO BURNED,
1200 PERSONS PERISHED
250 MILES AWAY

The evening of October 7, 1871, might well be termed the most tragic time in American fire history. The story of the Great Chicago Fire, started when Mrs. O'Leary's cow kicked over a lantern, is well known. Much of Chicago was leveled, with around 300 persons losing their lives. But a lesser known fire that began at the very same time on the very same evening about 250 miles away took the lives of 1200 people — more than four times the death toll in Chicago!

The pine forests surrounding the little town of Peshtigo, population about 1700, in Wisconsin's North Woods were dry. As in Chicago, no rain had fallen for over three months, and a strong wind was blowing through the trees. The day before, the local newspaper had declared, "Unless we have rain soon, a conflagration may destroy this town." Unfortunately, what the newspaper had feared indeed came to pass the next day.

Only those villagers who jumped into the Peshtigo River or sought safety along its banks survived.

At the same time fire alarm box 342 was struck in Chicago, 250 miles to the north an ominous red glow over the forest sent chills of terror into the hearts of Peshtigo residents, most of whom were doomed to perish within the hour. One of the survivors wrote:

> *In less than five minutes there was fire everywhere. The atmosphere quickly grew unbearably warm, and the town was enveloped by a rush of air as hot as though it were issuing from a blast furnace. The wind lifted the roofs off houses, toppled chimneys and showered the town with hot sand and live coals. The cries of the men, women, and children were scarcely audible above the rumble of exploding gas and crashing timber. People were numb with terror, seeing nothing but fire overhead and all around them.*

The only fire engine in Peshtigo, a hand operated pumper, was overwhelmed, and bucket brigade members soon fled the onrushing tongues of flame. Hundreds of crazed residents were running and milling about in panic, but they could not outrun the flames. Only those who jumped into the now warm water of the Peshtigo River or sought refuge along its marshy east bank survived. Within an hour every building in Peshtigo was leveled, as were the buildings in several other nearby towns and on some 400 farms.

Although word of the Chicago fire soon reached the nation by telegraph, it was not until three days later that news of the Peshtigo fire got even as far as the state capital at Madison. Relief trains, ready to be sent to Chicago from Madison, were immediately diverted to Peshtigo.

Today the Peshtigo Fire Museum stands as a tribute to those who lost their lives in the most destructive forest fire in history. By a strange coincidence, this fire and the Chicago fire, two of the worst fire disasters of all time in our nation's history, occured just 250 miles from each other, and, at the very same time!

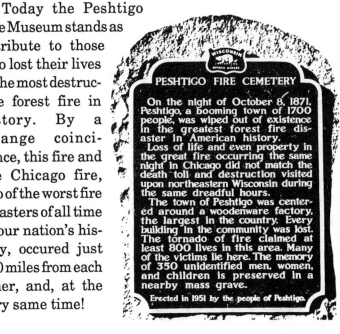

PESHTIGO FIRE CEMETERY

On the night of October 8, 1871, Peshtigo, a booming town of 1700 people, was wiped out of existence in the greatest forest fire disaster in American history.
Loss of life and even property in the great fire occurring the same night in Chicago did not match the death toll and destruction visited upon northeastern Wisconsin during the same dreadful hours.
The town of Peshtigo was centered around a woodenware factory, the largest in the country. Every building in the community was lost. The tornado of fire claimed at least 800 lives in this area. Many of the victims lie here. The memory of 350 unidentified men, women, and children is preserved in a nearby mass grave.
Erected in 1951 by the people of Peshtigo.

THE ALARM NEVER SOUNDED

AS NEW ORLEANS BURNED

William Shakespeare wrote, "A little fire is quickly trodden out, which being suffered, rivers cannot quench." Delayed alarms, as in the Great Chicago Fire of 1871, have resulted in horrendous loss of life and property. But, a century earlier, one of America's prominent cities was destroyed when the alarm never sounded at all!

In 1788, New Orleans, Louisiana, not yet a part of the United States, was governed by Spain and had a population of five thousand-three hundred and thirty-eight. The fires in the city were extinguished by bucket brigades and several hand engines which were summoned, as in most cities of the time, by the ringing of church bells. Can you imagine New Orleans burning to the ground — a total of 856 buildings — without a single toll on a single church bell? It really happened because the bells had been *forbidden* to be rung on the day the fire happened.

Almost all the residents of New Orleans were members of the Capuchin Order, and, in fact, the teaching of any other religion in the city was forbidden by law. The religious doctrine of the Capuchins, a branch of the Franciscans, forbade any ringing of church bells on Good Friday. Ironically it was candles on an altar which started the fire on March 21st, 1788, which was — you guessed it — Good Friday.

As the fire spread, nearby citizens cried out and, then, begged for the bells to be rung, which would have called the fire engines and bucket brigades. But, the Capuchin priests steadfastly refused. By the time runners had spread the alarm and the engines finally had arrived, the fire was far beyond their control.

It spread from house to house, from building to building, and within five hours nearly ninety percent of New Orleans lay in ruins: homes, stores, shops, the town hall, municipal building, jail, and hospital. What about the Capuchin priests and monks who refused to ring the bells? They watched with helpless dismay as their churches and chapels, as well as their monastery, burned to the ground.

The Forgotten Fire Engines

That Extinguished

Eight Out Of Ten Fires

They were the fast attack mini-pumpers of yesteryear. Whether hand-drawn, horse-drawn, or motorized, they arrived at fires fast, went into action within seconds, and had the fire out in a jiffy — if it was small. Yet these amazing engines had no pumps. The soda water in their tanks was forced through the hose and nozzle by pressurizing the tanks after they had arrived at the fire, by adding sulphuric acid.

Ever put your thumb over the top of a bottle of soda pop and shake it? You get instant pressure by releasing carbon dioxide from the carbonation in the pop. Instead of shaking the tank (usually 35 to 80 gallons) of soda water, yesterday's firemen released sulphuric acid from a glass bottle inside the tank, and bingo — carbon dioxide was generated to pressure the tank to about 175 psi. An instant chemical reaction did the trick.

❄ AMERICAN-LAFRANCE FIRE FIGHTING EQUIPMENT ❄

CHEMICAL FIRE APPARATUS

The records of the National Board of Fire Underwriters show that 80 per cent. of our fires are extinguished by the use of chemicals.

The remarkable success of chemical apparatus has led to an extensive use of this means of fire protection. Garages, stores, factories, industrial plants, private estates, rural communities—all find the chemical engine an indispensable safeguard.

The merits of chemical fire apparatus may be summed up briefly: This apparatus takes up little space and is always ready for immediate service; it is light, easily moved, and gets into service quickly; it requires but one or two men to handle; it is so simple any man of ordinary intelligence can understand it; it extinguishes fires with dispatch; it is particularly efficient on oil fires where plain water is almost useless; it saves water damage, which is often greater than the actual loss from fire.

The cost of maintenance of chemical apparatus is very small, as practically the only item of expense is the chemicals, which are commercial commodities easily obtained.

For these reasons a chemical engine in its various modifications has become an indispensable factor in every properly equipped fire department, and an absolute necessity in small towns and villages where the water facilities are limited.

An up-to-date modern factory is not complete unless one or more of these chemical engines is part of the equipment.

We guarantee advantages to purchasers that cannot be obtained elsewhere. We agree that all material and workmanship shall be of the best character obtainable, and will, at our own expense, replace such parts as may fail, where failure is attributable to defective material or inferior workmanship; and we agree that the apparatus and equipment will perform efficient duty, accident excepted, when properly and fairly handled.

An American LaFrance catalog page.

FIRE!!!

Instant Response

AMERICAN-LAFRANCE CHEMICAL *On Standard Ford Chassis*

THE safety of your own dear ones, your responsibility as a good citizen, demand that you get complete information about this wonderful motor fire-fighter at once. Be up-to-date and SAFE. "Chemical" has 40 times the fire-fighting efficiency of water. The outfit herewith is the product of the world's greatest manufacturer of fire apparatus.

The Saturday Evening Post, February 17, 1917

For half a century — from the early 1870's into the early 1930's — these chemical engines extinguished up to 80% of *all* fires simply because they arrived quickly and went into action immediately. No waiting to fill the tank from draft or with buckets. No waiting to get up steam. No waiting while a hydrant connection was made; they were, in effect, huge fire extinguishers on wheels.

Many of the chemical engines were constructed of brightly polished brass and copper tanks and fittings and were often adorned with fancy lamps and ornamentation, such as large eagle finials. So efficient were they that the firemen who operated them ascribed to them extraordinary fire-killing powers that they really didn't possess. Backed by unfounded statements from the National Board of Fire Underwriters, firemen firmly believed that the soda water in the tanks of their engines, when bolstered by carbon dioxide, could extinguish fires up to 30 times more efficiently than plain water.

Although this myth or legend persisted throughout the entire half century that chemical engines were in use, the soda water, even with the chemically-produced carbon dioxide, was, in fact, no more efficient than plain

The first chemical fire engine in the United States, sent over from Paris, France by the inventor of the soda-acid principal for extinguishing fires. This engine was the prototype for every chemical fire engine in America.

water. It was only the quick use of these engines on small fires that worked the magic.

In 1913, the Ahrens-Fox Company of Cincinnati, Ohio, one of America's foremost fire engine builders, came up with a new fast-attack system that spelled the beginning of the end for chemical engines. They introduced the "booster tank," which had a small pump to supply the pressure. No longer were chemicals needed to force the water from the tank. The changeover from chemicals to the booster system took another 20 years to complete, but by the time the last bottle of sulphuric acid had been poured into the last tank of soda water, firemen realized that the magic of the chemicals had been only an illusion — that plain water in the booster tank put out fires just as fast.

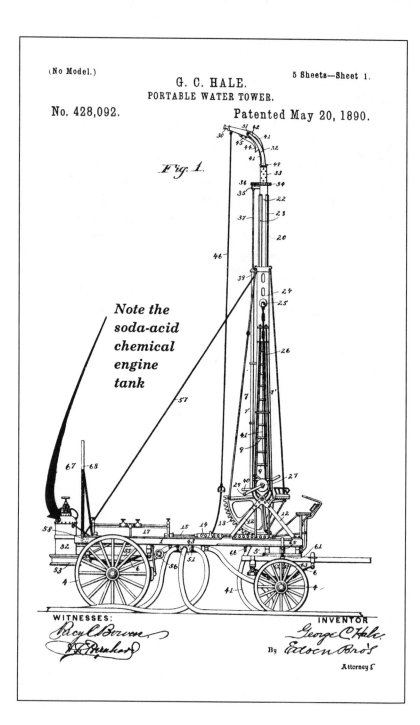

(No Model.)

5 Sheets—Sheet 1.

G. C. HALE.
PORTABLE WATER TOWER.

No. 428,092.

Patented May 20, 1890.

Fig. 1.

Note the soda-acid chemical engine tank

WITNESSES:

INVENTOR

By

Attorney

STILL ANOTHER USE

FOR "CHEMICALS"

In the late 1800's, buildings were getting taller, as the invention of the elevator eliminated climbing the stairs. But getting large streams of water into the upper floors of burning buildings was a real problem. The strongest aerial ladders could not withstand the pressure and weight of up to 1000 gpm at the top. Again, American ingenuity filled the gap, and the water tower was invented in 1878 by Abner Greenleaf, whose towers could blast water as high as the ninth floor.

Of course, the towers, pulled by horses, were folded flat until they were raised after arriving at the fire. But raising these heavy, unwieldy monsters was no easy task. Firemen struggled with hand cranks attached to gear assembles until Kansas City, Missouri, fire chief George C. Hale got a bright idea. His idea worked, and raising the heavy water towers became a breeze. In fact, Chief Hale went into the water tower business, both building them himself, and licensing his patent to other manufacturers. Here was his secret:

Chief Hale took the chemical tank (containing soda water with a glass bottle of sulphuric acid ready to break) and mounted it on a water tower, not to fight fire, but to raise the tower. The chemical reaction in the tank when the acid combined with the soda water, almost instantly produced up to 200 psi, and that pressure was channeled to two hydraulic cylinders which easily raised the tower. The first Hale chemically-raised water tower was manufactured in 1889, and the last one was built by licensee American-LaFrance 20 years later, in 1909.

Although water towers raised by other means were manufactured until 1937, they were used in large cities well into the 1950's and 60's, and in Memphis, Tennessee, until 1971. Water towers were supplanted, however, by still another type of fire apparatus, which you will learn about in the next story.

HALE ı PATENT ı WATER ı TOWER.

It consists of a strong oak frame-work, mounted on wheels, carrying an iron frame with an extending telescopic tube, through which passes the hose, conducting the water from the supply to and through the pipe on the top end. The motor or lifting power is furnished by a chemical tank.

IN DETAIL.

THE FRAME is of well seasoned oak timber, four rails firmly fastened and bolted together, about 4 x 6 inches, 19 feet long, and re-inforced by plates of iron, ¼ x 5 inches. The frame, when put together, is about four feet wide. This constitutes the frame or body of the apparatus upon which rests the tower proper.

TANK, CYLINDERS, IRON SUPPORTS, ETC.

THE WHEELS are of the Archibald patent, of sufficient size consistent with strength, and capable of bearing the super-incumbent weight. The axles are of steel of the Concord pattern, and the whole is carried on platform springs.

THE TANK is the ordinary chemical one as used on chemical engines, of the Babcock or Halloway pattern, and develops a rising power of from 10,000 to 20,000 pounds. It is simple in its operation and action, and the power generated therein is conveyed at will into brass cylinders and thence to long piston rods attached to the quadrant.

THE QUADRANTS work in a movable rack of wrought iron attached to the piston rods, and, being fitted on the shaft of the tower, by their turning raise the tower into position.

(Style 226.)

"HALE" PATENT WATER TOWER (without Deck-Turret.)

(Style 226.)

The Hale Tower is elevated by the means of chemicals that are stored in a cylinder, which cylinder is nearly filled with water. When it is desired to place a tower in operation at a fire, the chemicals are mixed, pressure produced, and by opening a valve the pressure is exerted on two pistons, which are meshed into cogs of a segment. By this pressure the piston rods are moved, thereby elevating the Water Tower.

This tower is now in use in the following cities :

New York City, N. Y.,	Milwaukee, Wis.,	Omaha, Neb.,
Boston, Mass.,	Syracuse, N. Y.,	Minneapols, Minn.,
San Francisco, Cal.,	Louisville, Ky.,	Cincinnati, O.,
St. Joseph, Mo.,	Philadelphia, Pa.,	Baltimore, Md.,
Denver, Colo.,	New Orleans, La.,	St. Louis, Mo.,
Buffalo, N. Y.,	Kansas City, Mo.,	Lowell, Mass.

WHAT?
FIGHT FIRE WITH A
CHERRY-PICKER?

In 1958, Chicago Fire Commissioner Robert J. Quinn found that his department had three antiquated water towers — a type of fire apparatus to throw heavy streams of water to the upper floors of buildings — but water towers weren't even being manufactured anymore.

Quinn had been watching tree-trimmers in Chicago's Parks Department using trucks with hydraulically operated elevating arms to lift them in baskets high into the air — the same type of apparatus used to pick fruit in orchards. Could these devices possibly be used to put large fire nozzles high in the air? Could they possibly replace his old water towers?

Conferring with the chief fire department engineer, he borrowed a Parks Department truck, strapped on hose and nozzle, and tried it out. It worked! Next Quinn checked with the Pitman Manufacturing Company of

A "Snorkel" or elevated platform is shown at work in its capacity of a water tower.

Grandview, Missouri, who helped his wild idea become a reality. In September of 1958, Pitman delivered a fifty foot elevated platform that was mounted on a General Motors Corporation chassis, and the Chicago Fire Department shop did the rest. They outfitted the rig with 3-1/2 inch diameter hose and a two inch diameter nozzle that could throw a walloping fire stream of 1,200 gallons per minute.

At 1:00 a.m. on October 18, Quinn's "Snorkel," the name given to the weird contraption by the news media, got a chance to show what it could do. It was dispatched to a four alarm lumberyard fire on Chicago's west side. With Firefighter John Windle operating the huge nozzle from the basket fifty feet in the air, the fire was extinguished in a fraction of the time that might have been expected. "It really plastered that fire in a hurry," and, "In 33 years of firefighting I've never seen anything as effective," were some of the comments by top Chicago fire officers.

Commissioner Robert J. Quinn soon became known as "Snorkel Bob" for bringing to birth one of the most versatile pieces of firefighting equipment ever devised. Today elevated platforms — some over 100 feet high — are used by hundreds of fire departments throughout the world.

INDIANAPOLIS 500 RACE CAR

DESIGNER'S INVENTION

REPLACED FIRE HORSES

It took three strong, well-trained horses to pull a full size steam fire engine in the early 1900's, and "steamers" were in use by most cities and towns of any consequence. They were rugged and dependable, and the municipalities had a great deal of money invested in them.

But after the turn of the century, fire apparatus manufacturers started to replace steam pumpers with engines that not only pumped the water, but were propelled to the fire by new-fangled internal combustion engines, powered by gasoline instead of horses. The new "motorized" fire engines were more efficient than even the old reliable steamers, but who could afford them? Few cities were willing to scrap their steamers and reinvest in motorized engines, even though they could replace and go faster than the faithful horses.

An enterprising designer of Indianapolis 500 race cars, John Walter Christie, came up with a solution. He

One of John Christie's tractors pulls Pittsburgh's Steamer 2.

formed the Front Drive Motor Company in Hoboken, New Jersey, in 1912, and proposed to pull steam fire engines to fires with gasoline-powered, two-wheeled tractors. His invention did the trick. During the seven year history of his company, he turned out more than 600 ninety-horsepower tractors, most of which pulled steamers. Some of them pulled ladder wagons and water towers as well. But as cities began to convert to motorized engines, the steamers with their horses and, then, even Christie's tractors began to fade from the scene. By the 1920's the vast majority of America's fire apparatus had become motorized.

Although other manufacturers "borrowed" Christie's idea and jumped on the bandwagon with their own gasoline-powered tractors, Christie's was the most popular. Whatever glory Christie failed to achieve on the track at the Indianapolis 500, he won with his valuable contribution to the fire service — the Christie front-wheel tractor.

40

Exhaust Whistles —

In Their Day

The Loudest Warning

During the decades of the 1920's and 30's, extra-loud mechanical sirens such as the Federal "Q," yelping electronic sirens, and compressed air horns were not yet on the scene, but a now almost-forgotten, audible warning device that was used on many fire engines could literally be heard for miles.

No other sound is comparable. It was an intermittent mixture of screeches, stutters, and trills. As the name implies, the exhaust whistle was simply mounted on the engine's exhaust pipe. In those days, mufflers were seldom used on fire apparatus, and the unmuffled roar of the motor with the strident,stuttering screeches of the exhaust whistle were many times as loud as the fire engine bells or the low-pitched mechanical sirens.

During those decades of exhaust whistle popularity, there was no air conditioning in homes and businesses, and screened windows were kept open during the summer. As a result, when fire apparatus went roaring and screeching to fires, their exhaust whistles were heard from one end of a medium sized town to the other.

The Buckeye company, perhaps best known in fire service history for the manufacture of Rotor-Ray warning lights, made most of the exhaust whistles. But just as the Federal Beacon Ray rapidly supplanted the Buckeye Rotor Ray, so did the Federal Q sirens replace the distinctive-sounding Buckeye exhaust whistles.

Anyone who has ever heard the raucaus screams of exhaust whistles on responding fire apparatus knows that no other sound today comes even close. The end of the exhaust whistle era brought down the curtain on one of the most unique sounds in American fire service history.

Two-Horsepower

Fire Engines

The earliest fire engines were pumped by hand. Next, steam powered the fire pumps, followed by internal combustion engines using gasoline and diesel fuel. But sandwiched in between the hand pumps and steam-driven pumps was an odd type of fire pump that was rated at two-horsepower because it was actually powered by two horses.

During the 1870's and 80's, these engines, although they never achieved much popularity, could be found in a number of fire departments. First, the horses pulled the engine to the fire; then they were hitched to the pump, which they powered by walking in a circle around the engine.

The most successful of these horsepowered engines was invented by fire apparatus pioneer Benjamin J.C. Howe, whose company continued to manufacture fire apparatus for another hundred years. Howe's engine was rated at 200 gpm and could throw a stream 160 feet through a 7/8 inch nozzle.

Howes Patent.

THE REMINGTON FIRE ENGINE admirably fills the gap between Hand and Steam Fire Engines, and combines the merits of both in effective work. It is a sweep power mounted on four wheels for transportation, and is operated by horses mainly, though it excels as a hand engine and can be turned around in its own length. It weighs about 3 000 lbs, being all metal except the levers and seat.

The axles, tires and spokes are wrought iron, hub of cast iron, diameter of rear wheels 40 inches, forward wheels 34 inches.

There are three double acting pumps so arranged and constructed that their combined action produces a continuous pressure and even flow of water, thus avoiding the vibrating motion as evinced by the ordinary piston pumps.

Diameter of the three double acting pumps, 5¾ inches ; length of stroke each, 8 inches ; capacity, 200 gallons per minute. The wearing surface is made of hardened polished brass to prevent friction and rusting.

We have a device for warming the pumps sufficient to prevent freezing in extreme cold weather.

When in use the engine is held in position by iron braces on each side, fastened to the ground by two iron or steel pins. One man can stake down while one or more men couple or attach the hose.

The drive wheel has eight spaces for levers to be attached, allowing twenty-four men to work the engine when necessary. We furnish two strong levers, so one or two spans of horses can be used, and they travel in a convenient circle at the ordinary walk of a work team, the draft being about the same as in plowing.

When operated by one pair of horses at such rate of speed as can be maintained by the hour, it will force through a 13-16 inch nozzle a stream 125 feet horizontally, or 60 feet in height and two streams through 11-16 inch nozzles, nearly the same distance. Of course two teams would make a large increase in volume of water, distance and height. In an emergency, where buildings are close together, or if it is inconvenient to reach water, the engine can be set where its width will permit it to run, and 2 or 3 feet motion to and fro of the levers by hand, will force a strong stream through the hose hundreds of feet distant.

The Engine is adapted to the standard 3½ inch suction hose, and 2½ inch discharge or leading hose.

We furnish with each engine the suction hose, wrenches, two hose pipes, each having different sized nozzles, lanterns, drag-rope reel, hand pole, and everything ready for operation, except the 2½ inch leading hose which we furnish with brass couplings at favorable prices, capable of 350 and 400 lbs pressure.

This Engine is always ready for work, and nearly as effective as a steamer, at less than one-third first cost, and not one-tenth the annual expense. It does not require the services of an engineer, fireman, mechanic, or other expert ; there is no waiting for a supply of fuel ; no time lost in getting up steam ; no danger from boiler explosion ; no flues to burn out, rust or blow up ; no expensive repairs. It combines economy, portability, and effectiveness, and can be quickly transported to a fire by men or horses taken where it would be impracticable to move a heavy machine, and put into operation immediately. The Engine is especially designed for villages, suburbs of cities, colonies, manufactories and large farms for is rigating pumping, &c. For the use of contractors, miners and others, in pumping out pits, mines, etc., and elevating water it is invaluable. When desired we can furnish hose carts, ladder truck and ladders, making a complete outfit at low prices. ADDRESS

Today there is only one of these "two-horsepowered" fire engines still existing. It is on display at the Firefighters Historical Museum in Erie, Pennsylvania.

FIRE!

Quick, Throw A Hand Grenade

During the early part of this century, one of the most popular fire extinguishers ever invented was in vogue. Although the vogue didn't last long, for awhile fire fighting hand grenades gave millions of home owners a false sense of security.

At the first sign of a fire, people just hurled a glass hand grenade at the flames. It shattered, and the chemical contents smothered the fire. A great theory, but unfortunately it didn't often work out that way. The glass grenades shattered all right, but the smothering capability of the chemicals, usually carbon tetrachloride, was ineffective except on a tiny fire in a confined space — an unlikely scenareo.

Some grenade-like extinguishers were not meant to be hurled but were mounted on walls and ceilings where the heat of a fire would break the glass, releasing the magic fluid, which was often colored to make it look more

imposing. These units were as useless as the grenades.

Of course the Underwriters Laboratories would not certify the grenades; and, after their popularity had run its course, they became flea market antiques. The concept of the grenades was neat, but they had one fatal drawback — they just didn't put out fires.

What Was
"Greek Fire?"

Before atomic missiles and bombs, before conventional missiles and bombs, even before gunpowder, back in the Middle Ages — a millennium ago — the most awesome weapon of war was Greek Fire; and it changed the course of history.

Greek fire, a substance made from Arab petroleum and a mysterious blend of chemicals, was hurled in clay pots like hand grenades or pumped from bellows-operated hoses or launched in flaming barrels from catapults. The napalm of the Middle Ages, it was used by the Crusaders, one of whom described it as follows:

It rolled forward as big as a barrel of wine, and the tongue of fire that shot out was as thick as a stout sword...It made such a noise as it flew that it seemed like thunder. It was as if a dragon were flying through the air.

The formula for Greek Fire was a closely guarded Byzantine state secret, thought to be a mixture of asphalt, niter, sulfur, and naphtha. It was used in both

A medieval knight charges with his Greek Fire lance - one of the most formidible weapons of its time.

land and sea warfare. At the siege of Jerusalem in 1099, the Crusaders victoriously entered the Holy City as the Saracens retreated before the flames of Greek Fire weapons.

Greek Fire was used extensively for five centuries in all of the Crusades. Not until the invention of gunpowder was Greek Fire superseded as the most awesome weapon of war. Its mysterious formula has been lost to posterity. Exactly how it was made will never again be known.

Note the Greek caption on this old drawing of Greek Fire in action. For centuries, prior to the invention of gunpowder, GreekFire was the most terrible weapon of war.

*"You can always preach an
illustrated sermon!"*

The Illustrated Sermon

During the early 1960's, near a prominent intersection in Perry Township, Vanderburgh County, Indiana, a ravine had been filled in by an enterprising religious group who proposed to pitch a gospel tent on the site. The "fill" had evidently included a considerable number of logs and tree limbs, which made air pockets under the surface.

How these subterranean logs and branches became ignited is unknown, but the volunteers answering the alarm were greeted by huge billows of smoke rolling up through cracks and fissures in the ground surrounding the tent.

The evangelist arrived shortly, and desperately pleaded, "Can you get it out? Our first meeting is tonight!" With that, one of the volunteers offered, "If we can't, Reverend, you can always preach an illustrated sermon!"

As thousands of gallons of water were pumped into the ground, the smoke gradually turned to steam and, then, subsided. The revival started on schedule.

The Story Behind The

"Fire Plug"

"Catch the plug!" the engine company officer shouts, and the fire engine stops momentarily at a fire hydrant as a fireman leaps off and wraps the end of the hose around the hydrant. The engine immediately proceeds to the fire, trailing a long snake of hose that falls from the bed of the truck, where it has been neatly folded.

That fire plug will soon be supplying hundreds of gallons of water each minute to fight a nearby fire. But what about the "plug?" What is it, and how did it get its name?

The fire hydrant or plug is, in fact, a huge water faucet. It is connected to a water main buried several feet under the ground — deep enough so that the water doesn't freeze on the coldest day of winter. The valve stem at the very top goes all the way down to the underground water main and opens the valve, letting water into the plug. Even in sub-zero weather, as long as the water is flowing, it will not freeze.

At Left: A fire hydrant or "plug" showing its below grade configuration.

Above: A hydrant or "plug" from the early 1800's.

54

The term "fire plug" harkens all the way back to about 1800 when the very first water mains in cities, such as New York and Philadelphia, were made of hollowed-out logs, with their tapered ends fitted together end to end.

To obtain water for firefighting from one of these crude mains, firemen dug down to the log, drilled a hole, and then kept their fire engine tanks full with the water that squirted out. After the fire was out, they — you guessed it — stopped up the hole with a wooden plug! And to this day — two hundred years later — a fire hydrant is more often than not referred to, by firemen and the public alike, as a fire *plug*.

THE REINDEER HOSE

COMPANY OUTING

The "outrageous" program of the Reindeer Hose Company (of Louisville, Kentucky) outing of Monday, August 14, 1893, was recently uncovered by fire historian David Winges, and is reproduced in its entirety without editing:

President Muscroft will preside with usual dignity, and after receiving Capt. Wm. Merker, of New Albany, with volunteers of said city, 150 strong, the following will be in order:

Part 1. — Capt. Merker will endeavor to deliver the Declaration of Independence in five different languages; this must be heard to be appreciated.

Part 2. — Capt. Hy Reamer, one of Albany's oldest veterans, and a Jiant in size, will, for the first time, attempt to eat more fish in a given time than Col. Muscroft's best record will show. The Albany boys will wager that he will win. He has lately eaten 8 lbs. of cat on trial, and knows what he is talking about.

Part 3. — Capt. Chas. Ochsenhirt will walk the tightrope, with Capt. Coyle on his back (blindfolded), and promises to accomplish this feat at the risk of Capt. Coyle's life.

Part 4. — Col. Adam Zimmerman, who has lately returned from the world's fair, with two live Laplanders will play freeze out, with any three members for the cigars; the 'Laps' are perfectly tame and harmless.

Part 5. — Capt. Tom Macdonald will sing, "When the Swallows Homeward Fly," in french and german; you must hear him, as it will amuse you very much.

Part 6. — Capt. Coyle will wrestle the big grizeley bear, two best out of three — catch as catch can — the bear will be muzzled on the occasion, to prevent bloodshed. No hugging allowed.

Part 7. — The Talking Match, between Capt. Zimmerman and Capt. Wm. H. Swift, will take place at 2:30 sharp; Adam says that he can do 765 words in one minute, while Capt. Swift claims 800 words. This will be strictly english.

Part 8. — Col. Jack Kaltenbach will attempt to drink two gallons of butter-milk in four minutes; something never attempted by any of the members.

Part 9. — Wm. Christopher Columbus Seibert will swallow a sword three feet long, and at the same time, a pint of pure cream. He has this trick down fine.

Part 10. — Capt. Chas. Pfeiffer, the old war horse, will sing, "Old Kentucky Home, " in German.

Part 11. — The foot ball tournament between the Albany Nine and the Reindeer Nine will take place at three o'clock sharp. This contest will be most exciting, as Capt. Ochsenhirt will play.

Part 12. — Col. Tom James and Col. Geo. Looms will have a sitting match, the one that gets up first will have to treat the crowd.

Part 13. — Capt. John Beha, the old Swiss veteran will show the boys how he can eat a pound of Swiss Cheese, and one Ib. of Smeer Kase in 15 minutes.

Part 14. — Col. Russell Cunningham, will, for the first time, dance the Highland Fling; in this he cannot be surpassed.

Part 15. — The old veteran, Davy Constance, will recite, the "Firemen of thirty years ago."

Part 16. — Capt. Coyle and Capt. Morningstar will dance the Schottisch, by request.

Part 17. — Capt. Wm. McNeil will attempt to jump 40 feet backward, if it kills him.

Part 18. — Mike Bocraft will wager that he can catch more fish in one hour, than any member on the grounds.

Part 19. — The Base Ball match, between the heavy and light weights will take place at 3:30 sharp. Capt. Westbay will referee, and no fouls will be allowed.

Part 20. — The Balloon Ascension by Profs. Frank Kochler and Jake Geopfert, will be the most interesting feature of the day; the balloon is the largest seen here, and brought here at a large expense. Capt. Oschenshirt will not go up in it, as he now weighs 345 lbs net.

The venerable Uncle Nick Haines and Dave Constance will preside at the Banquet, as they are both over 100 years of age, will be certainly interesting to our visitors. They remember well when Geo. Washington was one of the Laddies going to the fire with them.

By special request Capt. Ben Johnson will shoot an apple off Buck Meyers head 100 yards off.

P. S. Cappa's Band has been engaged for the occasion, at a fabulous price.

DECEMBER 7, 1941:

FIREFIGHTING AT

PEARL HARBOR

It was a peaceful Sunday morning in Honolulu when low-flying airplanes with Rising Sun insignias on their wings roared in just above the treetops at Pearl Harbor Naval Base. Not until a half hour after the original attack — at 8:26 a.m. — was the first alarm received at Honolulu fire headquarters. It was a call for mutual aid to Hickham Field, the nearby army base, the largest in the Pacific.

Arriving Honolulu firemen found the Hickham fire station in flames, both fire trucks destroyed, and the water mains blasted and useless. They laid over a mile of hose and drafted water from a bomb crater.

Soon the Honolulu fire alarm office was flooded with both box and telephone alarms. The department operated with only eight, two-piece engine companies (1000 gpm Seagrave pumpers and hose wagons) and a single 75 foot aerial truck. With three of the engine companies

The battleship USS Arizona, battered by bombs and torpedos which blew up its powder magazine during the sneak attack on Pearl Harbor, sank in less than nine minutes. More than 1000 of her officers and crew were killed. The U.S. Navy yard tug and fireboat, Hoga, battled fires aboard the Arizona for 48 hours. USS Arizona Memorial, National Park Service.

dispatched to Hickham Field, there were only five engines left for the entire city of Honolulu, but Fire Chief Bill Blaisdell knew how to improvise.

First he split the five remaining two-piece engine companies into ten one-piece companies by loading hose onto the pumpers and ordering the hose wagons to work directly from hydrants, using only the pressure in the water mains. Next he commandeered nine commercial trucks, loaded each one with 500 feet of hose, a nozzle, and a hydrant wrench, and used city and county employees, as well as volunteers, to man the makeshift apparatus.

One lone engine fought a fire at the Honolulu Gas Company which ordinarily would have been a multiple alarm. They controlled it in an hour and a half. Another fire ended up destroying 13 buildings and was finally stopped by hydrant streams from hose wagons — no engines were available.

A Japanese plane that crashed and burned was extinguished by a single fire engine, whose crew brought the dead pilot's charred body back to the fire station.

Of the three engine companies originally dispatched on the mutual aid call to Hickham Field, with 29 officers and men aboard, only 26 of them returned. As they bravely fought the flames consuming hangers, barracks, and planes on the ground, the second wave of Japanese planes came in, once again close to the ground, with machine guns spitting bullets and demolition bombs releasing.

Three Honolulu firemen, Captain John Carreira, Captain Thomas S. Macy, and Hoseman Harry T.L. Pang likely never knew what hit them as they made the supreme sacrifice in giving their lives for their country.

The Fireboat That Kept
The Pearl Harbor
Entrance Open

One of about 100 U.S. Navy vessels in Pearl Harbor on Sunday morning December 7, 1941, that comprised half of the Pacific Fleet was the combination yard tug and fireboat *Hoga*. Less than one year old, the 350 ton, 100 foot-long *Hoga* was rated at 4,000 gpm at 150 psi. She was berthed just a few hundred yards across the channel from "Battleship Row," which included the USS *Arizona* and USS *Nevada*. Battleship Row was number one on the Japanese hit list, and hit it they did.

When the action started, the ten man crew of the *Hoga* awoke to the screams of torpedo bombers whose wings were emblazoned with Rising Sun emblems, boring into Battleship Row. One explosion after another sent smoke and flames towering into the air as not only torpedos but also conventional bombs from other enemy aircraft found their targets.

The *Arizona* was hit repeatedly and eventually exploded when a 1,760 pound, 16 inch armor-piercing bomb penetrated to her forward powder magazine.

Within nine minutes she was on the bottom of the harbor in 38 feet of water, entombing 1,100 sailors and Marines. Soon joining her on the bottom were the USS *California*, the USS *West Virginia*, and the USS *Utah*.

Perhaps the two luckiest strokes of fate in this unlucky, nefarious, and unprovoked attack were, first, the absence of the three U.S. aircraft carriers that the Japanese were after, which their faulty intelligence had reported were berthed on Battleship Row. Had they been where the Japs expected to find them, the war in the Pacific would have been drastically changed. Second, the Japs overlooked the nearby gigantic fuel storage depot which contained the entire fuel supply for the Pacific Fleet. It went unscathed.

The Battleship *Nevada*, although hit repeatedly, remained afloat, as the *Hoga* sped from one doomed ship to another rescuing as many crew members as they could. The *Nevada*, although burning, headed down the channel for the narrow harbor opening that would lead to the open sea. But it soon became apparent that she would not make it. Fires were raging below deck, and she was sinking by the bow. If she went down in the narrow channel, the entrance (and exit) for the Pacific Fleet would be blocked.

The Japanese, realizing they could bottle up the U.S. Fleet, continued to pound the *Nevada* with more bombs. But the crew of the *Hoga* wasn't about to let that happen. Disregarding the rain of bombs, the *Hoga* pulled along side the *Nevada*, hit the fire with the monitor on top of the pilot house, and then actually boarded the doomed *Nevada* with four hose lines.

But as it became obvious that these firefighting efforts were too little too late and that the *Nevada* was

The Hoga, a U.S. Navy combination yard tug and fireboat, rushed to the aid of the USS Nevada after the battleship was severely damaged and set afire during the Pearl Harbor attack. Mooring to the port bow of the Nevada, the Hoga's pilothouse monitor gun bored into the flames while the tug's fire pumps fed hoselines put aboard the battleship. The Hoga pushed the Nevada across the channel leading to open sea and helped to beach her at Waipio Point to avoid the threat that the sinking Nevada would block the entrance to Pearl Harbor. USS Arizona Memorial, National Park Service

The Hoga was loaned by the U.S. Navy to the Port of Oakland after World War II. The combination yard tug and fireboat was converted into a single-purpose fireboat. While the boat retained most of its original configuration, its pumping capacity was raised from 4000 to 10,000 gpm at 150 psi.

about to sink right in the entrance to Pearl Harbor, the *Hoga* began to bravely push the battleship, six times its size, toward the shore, where she grounded on the bank of a sugarcane field. During the trip the *Hoga* kept the deluge from the pilot house monitor boring into the flames. It was "pump and roll" on an undreamed of scale.

The crew of the *Hoga* pulled it off — one of the most amazing navigational feats of the war. The *Hoga* finally pushed the *Nevada* into open water, where she grounded near Waipio Point. The entrance to Pearl Harbor remained open. The *Hoga's* crew was cited by Admiral

Chester Nimitz, Commander-In-Chief of the Pacific Fleet, for distinguished service with disregard for their own personal safety.

Within about two months after the surprise attack, the *Nevada* was refloated and towed to the Puget Sound Navy Yard at Bremerton, Washington, where she was repaired so she could rejoin the fleet and inflict revenge on the enemy who, but for the *Hoga*, would have sunk her. She participated in action at Iwo Jima, Okinawa, the invasion of France, and later, at war's end, in the occupation of Japan.

But whatever became of the *Hoga*? After the war she was upgraded from 4,000 gpm to 10,000 gpm and became the first fireboat for Oakland, California. Renamed the *City of Oakland*, she protected that city's waterfront until 1993, when she was finally taken out of service, almost a half century after the end of the War.

THE MOSQUITO ARMADA

During World War II, the London fireboat *Massey Shaw* (named for a London fire chief) fought hundreds of fires during the bombing of London between September 1939 and May 1941. But on Thursday, May 30, 1940, she answered the strangest alarm of her career.

She was dispatched to Dunkirk, France, some 40 miles from the tip of England across the English Channel, and about 110 miles from London. Her crew naturally assumed she was being sent to fight the waterfront fires blazing at Dunkirk, which included a burning fuel tank farm, so they loaded extra supplies of foam. But they needn't have bothered — their mission was not one of firefighting, for they were on their way to join Winston Churchill's Mosquito Armada.

At that particular time, the war in Europe was not going well for the Allies, and Churchill implemented Operation Dynamo — the code name for the evacuation of more than 300,000 British and French troops from Dunkirk.

A flotilla consisting of every available shallow-draft vessel, whether military or civilian, was assembled to accomplish the evacuation. This Mosquito Armada, so termed because of the comparative size of the tiny boats involved, included tugs, trawlers, motorboats, private yachts, and one fireboat — the *Massey Shaw*. The shallow draft of these little vessels enabled them to get close to the harbor and beaches at Dunkirk, and the *Massey Shaw* had been designed for the shallow inlets of London's Thames River. She was only 78 feet long with a draft of but three feet and nine inches.

On that Thursday afternoon, thirteen London firefighters volunteered to take her across the turbulent waters of the English Channel, a voyage she was never designed to undertake. She didn't even have a compass, so someone hurriedly bought a cheap one at a nearby shop. About 4 p.m., some two hours after having received the unusual "alarm," the *Massey Shaw* left London for Ramsgate at the tip of England, across the English Channel from Dunkirk, France. At Ramsgate, gray paint was slathered as camouflage over her brightly polished brasswork and white cabin,and her cabin windows were boarded up to keep out the waves she was totally unused to.

As the *Massey Shaw* approached Dunkirk, the crew had to maneuver between many of the 240 Allied ships already sunk in the shallow but heavy surf. They dragged and lifted soldiers on board, sometimes even jumping overboard to keep them from drowning. The *Massey Shaw* not only ferried about 500 soldiers, many of them wounded, out to a paddle wheel steamer, but actually made three, 80 mile round trips, with evacuees from Dunkirk, to Ramsgate. Total number of soldiers transported to safety across the channel was 646.

Of the approximately 400 boats in the Mosquito Armada the only non-military craft to be cited by Vice

The London Fire Brigade's Massey Shaw was the only fireboat among what Winston Churchill called The Mosquito Armada of little ships which participated in the World War II evacuation of troops from the harbor and beaches around Dunkirk. The Massey Shaw's all-volunteer crew was credited with rescuing 646 military personnel, including 37 survivors of a French freighter which sank after hitting a mine. Massey Shaw and Marine Vessels Preservation Society, Ltd.

Admiral Sir Bertram Ramsey, with the approval of His Majesty, King George VI, was the *Massey Shaw*, which received the Naval Distinguished Service Medal. Never before had this medal been awarded to firefighters.

The *Massey Shaw* remained in service with the London Fire Brigade until 1971, and she is now an historic landmark and tourist attraction at London's East India Docks at the South Quay Waterside Development.

WORLD WAR II
BAT BOMBS

Perhaps the most bizarre "secret weapon" of World War II, although never actually used against the enemy, was the plan to attach tiny incendiary devices to hundreds of thousands of bats, which would be released over Japan's combustible cities.

The incendiary mechanisms, weighing only an ounce, would not ignite until the bats had time to find roosts, such as belfries and under the eves of roofs.

Thousands of bats were collected from caves in New Mexico, and the miniature incendiary bombs were banded onto them. To test the bats' endurance at high altitudes, a plane load of them took off from a small airfield in New Mexico. Of course the project was top secret, with only sev-

A World War II "Batbomber" is shown flying with its one-oounce incendiary bomb which could give off a 22 inch flame for 8 minutes.

75

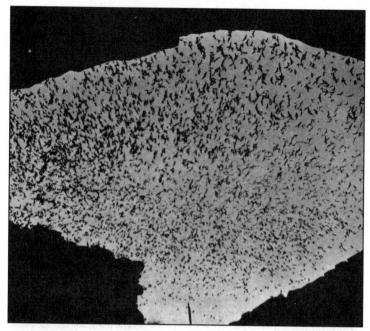

A nightly flight of thousands of Mexican freetail bats leaving Carlsbad Caverns, New Mexico.

eral persons aware of the reason for the flight. So, when some of the bats escaped, the cause of the fires that burned down the hangars at the small airfield was listed as unknown.

Had the war continued, rather than having been stopped by the Japanese surrender following the dropping of the atomic bombs, it is quite possible that the secret weapon of the incendiary bat bombs would have been used to burn many Japanese cities to the ground.

How Does
Fire Prevention In The
United States Compare?

Fire prevention practices in other countries are sometimes more intense and stringent than in our own. Here are some examples:

Japan — A person can be imprisioned for life for causing a severe fire by "grave negligence." At work, the Japanese learn to use fire extinguishers, and in some cities, wardens walk the streets to remind people to douse fires.

Netherlands — The building codes and insurance laws say that every room must have two exits.

Hong Kong — Apartment buildings appoint fire marshals.

Sweden — The fire academies train chimney sweeps to inspect fireplaces and furnaces, preventing dangerous fire starters.

England — The London Fire Brigade spends about a million dollars each year on fire safety commercials.

Korea — Seoul's equivalent of Times Square has 30-foot signs touting fire safety. Neighborhood fire drills are also conducted.

France — Insurance companies cover only partial costs of damage caused by fires in an effort to deter landlord arson.

Switzerland — Insurance companies pay only if an identical replacement structure is built.

THE INTERNATIONAL
FIRE PUMPING CONTEST
FIASCO

In 1863, the Prince of Wales, after witnessing an elaborate firemen's parade in his honor in New York City, was so impressed that he invited the firemen to send their best and most powerful engine to the Crystal Palace in London, England, for the international fire engine pumping contest.

The first successful steam fire engine had been invented only eleven years earlier. New York was not one of the first fire departments to buy one since, for four years, the volunteer firemen had rejected steam in favor of their cherished hand engines, but in 1856, the volunteers accepted the gift of their first steamer, the "Manhattan," from fire insurance companies who hoped it would reduce fire losses. It did.

The Manhattan was hand-drawn, lightweight (compared to most steam fire engines), and had brightly polished ornamentation. Soon the volunteers realized it

The Prince of Wales is on the balcony overlooking the 1863 New York City Firemen's parade held in his honor. He was so impressed that he invited the firemen to London for the International Fire Engine Pumping Contest.

was superior to their hand engines, and three years later they ordered two more. By 1863, the year of the contest, there were eleven steamers in New York, but the volunteers chose to enter the original Manhattan in the international competition. It was their favorite. Under the command of Foreman Charles Nichols she was loaded onto a ship which set sail for England.

Upon docking at London, the Manhattan, according to London firefighters, had to take a specific route to the Crystal Palace, and that route was up a steep hill. Although the London firefighters helped the Americans pull the Manhattan up the hill, once it had reached the top, they dropped back, leaving the Americans to control the engine's descent to the bottom of the other side of the hill. Because the 5,600 pound Manhattan had no brakes, it gathered more momentum than the New York firemen could control. Down the hill it rumbled, leaving the firemen behind. Its downward rush was stopped when the grand old streamer crashed into a tree, which knocked off the front carriage, broke the flywheel, and otherwise damaged the engine, to the extent that it was unsafe to pump.

Loath to admit defeat on English soil, the New York firemen, working at a furious pace, made temporary repairs, and the Manhattan entered the contest after all. As the contest got under way, the Manhattan quickly took the lead in pumping water up an incline. It appeared to be a sure winner when, suddenly, the damaged flywheel gave out. To add insult to injury, the London *Times*, in reporting the contest results, stated:

It must be understood that the American Steam Fire Engines are as much behind the steam fire engines of other countries, as that most pretentious Political Association called the "New York Fire Brigade" is behind any fire brigade in Europe in real usefulness.

What a slam! Were the Americans "set up?" Foreman Nichols and his men thought so, but once back home, they claimed a moral victory and celebrated with a parade and banquet.

THE FLAMING
SUNDAY SCHOOL PICNIC
DISASTER

How could more than 1,000 persons, most of them children, lose their lives at a Sunday School picnic? This bizarre tragedy, one of the worst in America's firefighting history, began to unfold on Wednesday morning June 15, 1904, when over a thousand happy members of Manhattan's German Lutheran Church Sunday School began the walk from the church on 6th Street to the dock at the foot of East 3rd Street.

Arriving at the dock, the picnicers crowded aboard the big side-paddlewheel excursion steamer *General Slocum* for a trip up the East River to the picnic grounds in the Bronx. As the 264 foot-long triple-decker excursion boat pulled away from the dock with 745 children, 563 women, and 50 men plus Captain William VanSchaick and his 22 man crew, a band on the hurricane deck struck up lively German music. As the boat's whistle tooted, they pulled out into the river bound for the picnic grounds at Locust Point. But the ill-fated vessel, now crowded with more than 1,400 persons, never made it.

Sixty-seven-year-old Captain VanSchaick was the only captain the thirteen-year-old *General Slocum* had ever had, and as he piloted the boat, the children laughed and played and sang with happiness. Then several of them on the lower deck noticed wisps of smoke seeping from cracks around the door of one of the cabins. When a crew member opened the door to investigate, he was greeted by flames flaring up as a result of the fresh air introduced into the cabin.

The crew had never had a fire drill nor received any firefighting training. But acting on instinct, they pulled a fire hose and opened the valve. No water came out. By the time the engineer started the fire pump and debris was cleaned out of the valve, the flaming cabin looked like a blast furnace. As water at last flowed through the linen fire hose, it promptly burst in several places because it was rotten. Still no water reached the nozzle.

A man on the shore saw the *Slocum* trailing smoke and immediately pulled a fire alarm box, supposing the captain would head for shore. But Captain VanSchaick keep right on going at full speed with the wind fanning the spreading flames. The crews of the responding fire engines could only watch with dismay as the flaming boat kept on steaming up the river. Next a fireboat was dispatched, which pursued the burning *Slocum,* but neither could it catch her. Captain VanSchaick was obviously not about to slow down nor head for shore. Hundreds of women and children leaped from the burning boat into the river, many of them with useless life preservers which disintegrated in the water. Made of

Postcard producers often capitalized upon major news events, especially when actual photos failed to capture the fire spectacle. This 1904 postcard appeared shortly after the General Slocum disaster on June 15 of that year. Except for the fairly good representation of The New Yorker, the blazing excursion steamer lacks the Slocum's side paddlewheels. Apparently to hype the scene, the artist chose to show the burning steamer under a full moon. The disaster occurred during a sunny morning. From the collection of Steven Lang

canvas, they were rotten with age, like the fire hose, and they were filled with only granulated cork.

Finally, VanSchaick beached the doomed vessel off Locust Island, where two fire boats caught up with her, and although their crews made gallant rescues, the loss of life was a staggering 1,030. Captain VanSchaick was arrested on the spot. Nine hundred and fifty-eight bodies were recovered, including the minister's wife and daughter, and were buried in the Lutheran Cemetery.

As an aftermath of the tragedy, President Theodore Roosevelt fired the steamboat inspectors, including the chief inspector who was eighty years old. Captain VanSchaick was tried on two counts of manslaughter and a third count of having failed to train his crew in firefighting procedures. He was found guilty on the third count and sentenced to ten years at hard labor. Except for his pardon seven years later by President William Howard Taft, he likely would have spent the rest of his life in Sing Sing prison.

Squire Boone's
Fire Extinguishers

Although fame eluded Daniel Boone's kid brother, Squire, his daring exploits were no less colorful and dramatic than those of his famous brother. Gunsmith, carpenter, builder and operator of grist mills, fearless explorer and Indian fighter, as well as being a part-time Baptist minister, Squire Boone (1744 — 1815) was a man of many talents. And those talents included making the fire extinguishers that helped save Fort Boonesborough (Kentucky) from what has gone down in pioneer history as "The Siege of Fort Boonesborough."

Of the first eight white men who dared to enter "the dark and bloody ground," as Kentucky was known to the pioneers in the early 1770's, only two returned alive — brothers Daniel and Squire Boone. But in 1778, during Squire's attempt to colonize Kentucky, he and 64 men, along with women and children, were trapped inside Fort Boonesborough when they were surrounded by 450 Indians who were after their scalps. Squire and his forters had food and water, but their greatest fears hinged on their scant supply of ammunition and the threat of fire. They knew, as did the Indians, that their fort was highly combustible.

SQUIRE BOONE

Indiana Historial Bureau
From the collection of Frederick P. Griffin

Photography by Mike Tonegawa

Squire Boone's original carvings in stone are on display in Boone's Mill at Squire Boone Village near Corydon, Indiana.

Drawing by Violet Windell

Squire, seriously wounded by Indian rifle fire, gave instructions from his bed of pain for constructing fire extinguishers — the first ones west of the Allegheny Mountains. Buckets of water would not only have expended the precious fluid which they needed in order to survive, but in throwing water up onto the rooftops of the fort, they would expose themselves to arrows and bullets.

Realizing that the Indians would likely attempt to set the fort on fire with flaming arrows, Squire came up with a novel plan to extinguish roof fires. Old rifle barrels were salvaged, wooden pistons made to fit inside the barrels, and the women and children were pressed into service as firefighters to squirt water up onto the roofs while they were in the fort below safe from the arrows and bullets.

The Indians did shoot flaming arrows, and Squire's extinguishers worked like a charm, putting out fires as fast as the burning arrows landed. Not only was the fort saved — the Indians finally gave up and disappeared into the forest.

Today visitors may tour recreated Fort Boonesborough near Winchester, Kentucky, as well as watch the miller grinding grain at Squire's recreated grist mill at Squire Boone Caverns and Village near Corydon, Indiana and Louisville, Kentucky. Without Squire Boone's makeshift fire extinguishers the history of Kentucky would have been a different story.

Squire Boone, and the other Fort Boonesborough representatives at the peace treaty, which was merely a treacherous rouse by the Indians, rush back to the fort. Squire was shot and nearly died from his wound. Daniel cut the bullet out of Squire's shoulder, as Squire held a broadax in readiness in case the Indians got inside the fort. From his bed he gave instructions for the making of the crude fire extinguishers which saved the fort from the Indians' flaming arrows.

Artist: Harold D. Collins

The building of Fort Boonesborough.

Artist: Russell May

Fort Boonesborough completed as viewed from the opposite bank of the Kentucky River.

Colonial

Fire Companies'

Unusual Tool

The equipment of early fire companies in the American colonies was limited to only several items: leather buckets, ladders, swabs for roof fires, axes, hooks for pulling down burning buildings, canvas bags for carrying household goods from burning dwellings, and the item pictured.

The most valued and prized possessions of early colonists were their beds. When a house caught fire it was expected to go to the ground, so the priority items to rescue were the beds. But the beds wouldn't fit through the doors. The metal bed frames were originally carried into the house in pieces and then assembled; so, to carry them out they had to be disassembled, which required a wrench. The bed key was a one-size-fits-all wrench, used to quickly take apart any bed for hasty removal from a burning home.

So, if you were a volunteer fireman in Colonial America and were rudely awakened by the cry of "Fire!" or a loud rattle (a noisemaking rachet device carried by night watchmen to arouse the volunteers) or the tolling of a bell, you would quickly don your clothes, grab your bucket, and run to the fire, making sure that in your pocket was your trusty bed key.

GAMEWELL -

THE FIRE ALARM TYCOON

He couldn't have done it today, but in the late 1800's anti-trust laws didn't exist. If, by whatever means, a company could gain a virtual monopoly on its products, there was no one to stop them. No one stopped John Nelson Gamewell, the entrepreneur who cleverly and ruthlessly cornered 95% of the world's municipal fire alarm market.

No one would have suspected that the man to pull off this amazing coup was a bookseller by trade, a village postmaster, and an amateur telegraph operator. Yet, this unassuming portly, balding 33-year-old telegraph hobbyist from Camden, South Carolina, was destined to have his name become synomous with the transmission of fire alarms.

William F. Channing, a medical doctor turned inventor, patented the first fire alarm telegraph system and gave a lecture on its merits at the Smithsonian Institution in 1855. Making the trip from South Carolina to hear Channing's spellbinding exhortation was

John Gamewell, an amateur telegrapher. He realized not only the significance of improved fire alarm transmission but also the vast amount of money to be realized for whoever could own and control the new technology.

A man of little means, Gamewell returned to Camden in a dither — how could he raise the money to buy Channing's patents? He convinced his friend, James Dunlap, a wealthy jeweler and merchant, to put up the money for the rights to the patents for the southern states. Within four years Gamewell had raised another $30,000 and had bought the patent rights for the entire country.

By 1861, Gamewell had sold and installed fire alarm telegraph systems in Philadelphia, St. Louis, Baltimore, New Orleans, and Charleston, South Carolina. However, the Civil War derailed the growth of the budding alarm empire. Gamewell served the Confederacy as an official of South Carolina's War Department, which included managing the saltpeter works near Columbia where gunpowder ingredients were manufactured. Then, one day during the war, General Sherman's troops neared Columbia, and Gamewell fled back to Camden, where he hid in a swamp.

After the war, Gamewell was not only back to the start of his endeavor, he was destitute. The U.S Government had confiscated his fire alarm patents and had auctioned them off from the steps of the Camden City Hall. Gamewell, totally without funds, borrowed five dollars from his brother-in-law, James Gardiner, who was a clockmaker, and who would, in due time, invent improvements to the clock-like mechanisms of fire alarm box movements.

JOHN NELSON GAMEWELL, 1822-1896

A South Carolina postmaster and telegraph company agent, John Nelson Gamewell founded the company whose name was to become synonymous with fire alarm boxes and fire alarm telegraph systems throughout the world. Although there are at least 36 other known manufacturers of fire alarm telegraph equipment, Gamewell held a virtual monopoly, with a 95% market share.

From the time he joined Gamewell in his business, Gardiner worked daily at the company until he died more than a half century later, at age 95.

The wealthy Dunlap refused to put up any more money, so Gamewell moved his wife and family to Hackensack, New Jersey. While there, he befriended a gentleman named John F. Kennard, who was somehow able to repurchase all of Gamewell's confiscated patents for a mere $80.00. With Kennard's backing, Gamewell set up a manufacturing facility in Upper Newton Falls, Massachusetts. The company thrived, and although competitive businesses sprang up, the shrewd Gamewell developed the largest and most successful one by far.

As the business continued to prosper, Gamewell either bought out his competitors or forced them out of business with the power of his growing alarm empire. By 1904, there were 764 fire alarm systems in the United States, with a total of 37,739 fire alarm boxes, of which 95% were Gamewell.

At the zenith of fire alarm systems in the United States, which was probably the two decades after World War II, there were nearly 2,000 Gamewell fire alarm systems in operation not only in this country but also throughout the world, and these systems contained nearly a quarter of a million fire alarm boxes — the familiar "red box on the corner."

As the percentage of false alarms from the boxes grew to alarming proportions, and as telephones proliferated, the expense of maintaining the Gamewell systems became a burden few cities were willing to continue. Only a handfull of communities and one large city — New York City — still have fire alarm telegraph systems. The old boxes, removed from street corners, are now collectors' items.

As for John Gamewell, the man with the vision for building a fire alarm empire — which in fact he did — he died of heart problems at his Hackensack home in 1896 at age 73. His death was little noted, but the Gamewell name always will be remembered for the red fire alarm boxes on America's street corners that signalled millions of fire alarms for over a century.

THE FIRE ALARM GENIUS

WHO BLEW

HIS BRAINS OUT

lthough fire alarm telegraphy was the brainchild of Dr. William F. Channing, a Harvard Medical School graduate, and although the father of fire alarm telegraphy is generally conceded to be John N. Gamewell, an obscure telegraph operator, turned financial wizard, who cornered the fire alarm business with a 95% market share, yet it was the mechanical genius, Moses Crane, who enabled fire alarm systems to really function.

The son of clockmaker Aaron Crane, Moses invented and patented mechanisms for bell-striking machines, clock-like movements for fire alarm boxes, and electro-mechanical fire station gongs. The enterprising John Gamewell soon realized that Crane's patents could be worth a fortune, so in 1886, he bought out all of them for $47,000 and had Crane sign a non-compete agreement for ten years.

One of Moses Crane's electro-mechanical fire house gongs.

Crane, however, couldn't stand idly by and watch Gamewell become a fire alarm millionaire. He had to try something. So, Crane kept on inventing fire alarm improvements, which were patented by his protégé Frederick Cole. Ultimately, Gamewell grew wise to the subterfuge and sued Moses Crane for breach of contract.

Although he was well off financially, Crane brooded over his shattered dream of major fire alarm success. Depressed by what his granddaughter termed "heartbreaking lawsuits and ruin," in 1898, at age 65 he put a pistol to his head and pulled the trigger. The inventive genius of Moses Crane, the man who perfected fire alarm telegraphy, had come to an end.

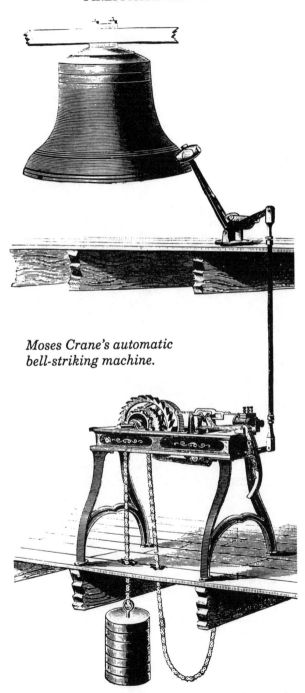

*Moses Crane's automatic
bell-striking machine.*

103

The "Improved"
Fire Alarm Box
That Backfired

Prior to 1875, fire alarm boxes were kept locked to prevent false alarms. Persons who turned in alarms had to secure the key from a nearby merchant, policeman, or trusted citizen. Needless to say, many alarms were delayed, resulting in loss of both property and life.

Entrepreneurship in America has always risen to fill a need, and Charles Tooker was equal to the challenge. In 1875, he patented a "keyless" fire alarm box which rang a loud bell when the door of the box was opened. This alerted passers-by that the box was being activated.

Gamewell, as might be expected, soon bought the patent, but foolproof, Tooker's box was not. Not only did the person turning in the alarm have to turn the handle and open the door of the box (thereby ringing the bell), he then had to pull the hook inside the door to transmit the alarm. The instructions were plainly printed on the front of the box in bold letters, but how many excited persons trying to turn in a fire alarm ever paused to read them? Just turn the big handle, the bell rings, and the fire department is on the way — right?

Wrong! The fire department doesn't have a clue that there is a fire until the hook is pulled, which many excited persons failed to realize. Tooker's improved keyless fire alarm boxes were installed on thousands of street corners in cities throughout America, but alas, the solution became a bigger problem than the false alarms they were designed to prevent. Finally the National Board of Fire Underwriters ordered all of the bells from Tooker's fire alarm boxes removed.

Today there are only a few remaining original Tooker boxes with the bells still intact, and they are rare collectors' items. At fire memorabilia auctions they sell for several hundred dollars each.

Will he really turn in the alarm by opening the door and pulling the hook, or will he just turn the handle, hear the bell, and assume the alarm went through? Many buildings and lives were lost because of this false assumption.

Prior to 1870, Fire Alarm Boxes Were Kept Locked!

Trying to locate the key from a policeman or merchant often resulted in delayed alarms, with property destroyed and lives lost.

The Mystery of
The Hoodoo Box

Buffalo, New York's fire alarm box number 29, at the corner of Seneca and Wells Streets, was a jinx. In fact, so many Buffalo firemen were injured and killed fighting fires reported from that box that firemen dreaded answering the alarm whenever box 29 tapped in. Its reputation as the "hoodoo box" was well deserved.

On July 18, 1878, flames leaped from the roof of the Red Jacket Hotel, and a wall collapsed killing Fireman John D. Mitchell. On February 2, 1889, two factories, two hotels, and the entire city block were destroyed. Fireman Richard Marion was killed when a wall collapsed on him. Fire Chief Fred Hornung had his arm almost severed by falling glass. On January 28, 1907, a collapsing wall at the Columbia Hotel fire buried 20 firemen, killing three of them. On July 19, 1919, several commercial buildings on Seneca Street burned, and 30 firemen were overcome by smoke.

Alarms for many other serious fires came in from box 29 over the years until the National Board of Fire Underwriters conducted an investigation. Their findings — the serious and fatal fires reported from box 29 were purely coincidental.

The last major fire reported from box 29, before all Buffalo fire alarm boxes were removed, was on December 30, 1981, at around 11:00 p.m. The fire was in an old four-story brick building. A tall man in a long, dark overcoat led the first-arriving firemen into the building and then to the seat of the fire. As the fire mushroomed up the elevator shaft and was spreading onto the upper floors, a second alarm was struck from box 29.

As the fire gained in intensity, all firemen were ordered out of the building, and the third alarm was transmitted from box 29. The fire raged all night, but by morning, it was finally under control. As fire inspectors began their search for the cause of the fire, they were told, by the men of Engine 1 and Ladder 2, about the tall man in the long, dark overcoat who had led them into the building and had directed them to the seat of the blaze. They had supposed that he was the night watchman.

However, the man had disappeared. The building owners said that they had no watchman and that the building had been locked when the last employee had left, hours before the fire was reported. The inspectors found the dead bolt on the front door in the locked position. But it was through that door that the mysterious man had led the firemen.

How did he get through the locked door? How did he manage to leave the building locked? Who was he? He is part of the mystery of the hoodoo box — a mystery that will likely never be solved.

For a number of years, during the annual Firemen's Ball in Buffalo, a period of silence was observed in memory of all Buffalo firemen who had died in the line of duty. Then the silence would be shattered when a fire department bell clanged out two strokes, a pause, then nine strokes — the number of the hoodoo box. Although the box is gone from the corner of Seneca and Wells, the legend lives on.

110

When Buffalo began removing its fire alarm boxes, Hoodoo Box 29 was preserved and mounted at the Buffalo Fire Department Museum. Although many Gamewell boxes had preceded it, the last Hoodoo Box was of a 1931 style with an exterior shell of lightweight Herculite and a white-painted quick action door. It was activated by pulling down the handle and then a lever. Paul Ditzel

Three Buffalo firefighters were killed while fighting a fire, January 28, 1907, which was sounded from Hoodoo Box 29. Paul Ditzel

WHERE'S THE FIRE?

Louisville, Kentucky, at the Falls of the Ohio River in the 1830's, utilized what is possibly the most unique method of directing volunteer firemen to the scene of a fire ever devised. In a tower atop the fire house of Mechanic Fire Co. No. 1 was a beautifully carved seven foot high wooden statue of a pointing fireman.

Various other cities of the time had fireman statues mounted on their stations, but this one was different — it swiveled to point in the direction of each blaze. Responding volunteers could glance at the roof of their station to see which way the "Chief Director" was pointing.

An iron pole extended from the roof tower housing the Director to the ground floor, where a wheel was affixed to the pole. Turning the wheel until the arrow on it pointed in the direction of the fire, aligned the pointing finger at the other end of the pole with the arrow. Thus, the firemen easily rotated the Director, pointing him in the proper direction when an alarm sounded.

The Director faithfully pointed the way to each fire until the volunteer department was disbanded in 1858,

Photo: *Caufield & Shook Collection.*
Photographic Archives, University of Louisville.

J.B. Speed Art Museum

*The Chief Director is still pointing the way to fires just as he did
more than one and one-half centuries ago. But now he points
from a position of honor in the J.B. Speed Art Museum in
Louisville, Kentucky, instead of from his original perch atop
the fire house of Mechanic Fire Co. No. 1.*

at which time Louisville could boast having the second all-paid fire department in the United States. (Cincinnati was the first.)

The Chief Director, no longer needed by the new department, was remounted on the Volunteer Firemen's Historical Association Building as a memento. When that building was razed in 1964, the Chief Director was moved to the J.B. Speed Art Museum in Louisville, where it may be seen to this day, still proudly pointing as in days of yore.

OOPS —

THE WRONG
FIRE HOSE COUPLINGS

As 34 steam fire engines began to arrive in Baltimore aboard railroad flatcars from other cities in response to a desperate call for help, no one had ever given any thought to a seemingly insignificant item — the threads on the fire hose couplings.

On February 7, 1904, the efficient and well-equipped Baltimore Fire Department responded to a fire alarm at the Hurst Building, a wholesale dry goods company in the commercial district. Arriving firemen found no evidence of a fire, but they checked out the building as standard procedure. Then, they discovered a small fire smoldering in a pile of trash in the basement, so they brought in the small hose from a chemical engine to put it out. But as they disturbed the pile of smoking

117

trash, they were greeted by what seemed to be a blast furnace. They dropped the small hose and ran for their lives.

Soon the fire blew out the windows on all six floors, and the entire Baltimore Fire Department was called — twenty-nine steamers, hose wagons, chemical engines, and eight ladder trucks. But they were not enough. The fire began to spread to other buildings.

Baltimore's thirty-six-year-old mayor Robert M. McLane sent calls for assistance to fire departments in Washington, D.C.; Wilmington, Delaware; Philadelphia, and New York City. Within a few hours they started to arrive, but although there was plenty of water in the mains, the out of town engines couldn't hook up to the fire hydrants because their hose threads didn't match.

They had to be content with pumping water directly from the Chesapeake Bay and relaying it all the way to the fire. In spite of their best efforts, Baltimore's central district was leveled — 1,343 buildings.

Although there was much criticism of the way the fire was handled, (Mayor McLane committed suicide within four months of the fire) two factors are obvious.

Upon discovering the fire in the basement, firemen should have stretched a 2-1/2 inch hose line to back up the small hose from the chemical engine, but more importantly, they should have had the forethought to have adapters on hand, thus enabling engines from other cities to hook up to their hydrants. For want of adapters, a city was lost.

Note: Today National Standard threads are almost universal on fire hose and hydrants throughout the United States.

THE EPIZOOTIC FIRE

have you ever had the epizootic? Although you may have kidded about it, you never really have had it, even though it's a real disease, because it's a disease only horses can get. What distemper is to dogs, epizootic is to horses.

If all the fire engines in Boston were pulled by horses, and all the fire horses got the epizootic, what would the results be? The following bit of fire lore tells exactly what happened when this situation really did exist:

In October 1872, the year after the Great Chicago Fire, an epidemic of epizootic spread from Montreal and Toronto in Canada to the United States. Within a day's time, 300 horses were dead in Buffalo. The disease spread on to Philadelphia, to New York, and then to Boston. By November fourth all 93 Boston fire horses were down with it. The heavy steamers, weighing at least three tons each, had to be pulled by hand.

On November 9th at 7:24 pm. box 52 at Sumner and Lincoln Streets in downtown Boston was pulled. One fire company's horses had recovered, and a single steamer was pulled to the fire by its own horses. Four other

Horse down! In October 1872, all 93 Boston fire horses were down with the epizootic .

companies borrowed horses, but they weren't strong enough to pull the heavy engines without resting on the way. The remaining 16 Boston steamers, as well as hose carts and ladder trucks were all pulled by hand — a tremendous test of endurance for the firemen, who finally arrived at the fire exhausted.

So much time was lost that the fire soon spread out of control. Chief Damrell called for help from every city and town within 50 miles. The response was tremendous, with 45 engines, 52 hose wagons, 3 ladder trucks, and 1,689 firemen arriving, most of them on railroad flatcars.

However, it was too late to save Boston. The delay in Boston's fire engines getting into action doomed the city. The fire raged for 16 hours, destroying 776 buildings, leaving 20,000 persons jobless, 1,000 homeless, $76 million in damage, and thirteen firemen dead, nine of them from other communities who had come in response to Chief Damrell's call for assistance.

Boston was destroyed because the fire horses were sick. No wonder a Boston district fire chief renamed the Great Boston Fire the Epizootic Fire.

Fighting Fire

With Fire

Do firemen ever really fight fire using fire? There are three scenarios where the answer is "yes" — one of them direct, and the other two indirect. The direct method is, of course, a backfire, when firefighters deliberately start a brush fire or forest fire to burn toward an existing fire. When the two fires meet, the fires have expended the available fuel and burn themselves out. This is often a last resort approach. The wind must be right, and the firefighters must know exactly what they are doing; still, it often works.

One of the indirect methods of fighting fire with fire was used during the era of steam fire engines — from the 1850's into the 1920's. Each steam fire engine had a boiler heated by a fire burning right on the engine itself. As in a steam railroad locomotive, a coal burning fire heated the water in the boiler to produce steam. In a railroad engine, the steam turned the locomotive's wheels to propel the train, but in the fire engine, the steam propelled the fire pump, which on the largest engines could pump up to 1200 gallons of water per minute.

A fireman uses a torch to start a backfire. Note the flames of a forest fire above the trees.

The sight of a steam fire engine racing down the street pulled by three horses, with black smoke belching from the stack, was a sight to behold and to be long remembered. The firemen were using the fire, right on the fire engine itself, to fight the fire, so to speak.

Still a third fighting "fire with fire" scenario evolved during World War II when the "indirect method" of fire attack was pioneered by Chief Lloyd Layman of the Parkersburg, West Virginia Fire Department, which method was also adopted by the United States Army and Navy. The fire in this case was the fire that needed to be extinguished, and the firemen used it to extinguish itself. Sound crazy? Here's how it worked, and still works today: steam is an excellent fire extinguisher. Why not let the fire generate steam to smother itself out?

This was accomplished by introducing tiny droplets of water — millions of them in the form of fog — into the layer of superheated air, gasses, and smoke that were trying to rise but which were trapped by a ceiling or roof. The smaller the droplets the better. (A mist as fine as water being sprayed from an atomizer is the most efficient). The small droplets of water sprayed into very hot air immediately vaporized into clouds of steam.

Using this method, firemen did not ventilate the building with the usual hole in the roof; but, they introduced water particles into the trapped layer of hot air and gasses, using a fog nozzle, getting unbelievable results in many cases. The fire was quickly blackened out by the resulting steam, with very little water damage resulting, since very little water was used.

The smoke in this picture is from the stack on the boiler of this steam fire engine. The fire under the boiler heats the water to produce steam, which powers the pump on the engine. Picture courtesy of Valdez Heritage Center, Valdez, Alaska.

This type of indirect attack became popular in the United States in the 1950's, 60's, and 70's, but it has lost favor in recent years. In several countries in Europe, and in New Zealand, the indirect method is still standard procedure whenever hot air and gasses that are trying to rise are held in place by a ceiling or roof. The heat from the fire and the fog generate steam, and the steam puts out the fire, which is literally fighting fire with fire.

An illustration from the first book ever written on the indirect method of attack. From Attacking and Extinguishing Interior Fires *by Chief Lloyd Layman, published by the National Fire Protection Association in 1952.*

THE TRAGIC
AERIAL LADDER
DEMONSTRATION

In 1866 Daniel Hayes, a delivery engineer, arrived in San Francisco with a fire engine and an idea. He was delivering a steam fire engine that had been manufactured in New Hampshire. Watching him testing the new engine, the San Francisco firemen were so impressed with his ability that they offered him the job of chief mechanic with their department. He jumped at the chance because here was the opportunity that would allow him to work on an idea he had.

Prior to that time, even the longest extension ladders (up to 65 or even 75 feet in length) had to be raised by hand, a back-breaking struggle involving many firemen. But, the innovative Hayes had the idea of attaching the base of a long ladder to a wagon and raising it mechanically.

By 1870, he had a prototype ready for testing, and it worked. He patented his mechanism and sold the rights to several manufacturers. However, an Italian firm

copied his idea and was able to patent their close copy. They sold the U.S. patent to Marie Belle Scott-Uda, who contacted the secretary of the New York Board of Fire Commissioners, William White. White struck a deal with her, agreeing to order three of her aerial ladder trucks if she would sell him the patent for sixteen thousand dollars. They closed the deal, and White brought in New York City's first aerial ladder truck for a public demonstration.

The truck was delivered to Rutgers Square, raised to its full height, and seven New York firemen began to climb. When the first three reach the fly or extension part of the ladder, it began to twist and buckle until it snapped in two, dropping all seven firemen to the pavement below before the horrified eyes of hundreds of spectators. All three firemen on the fly ladder (one was a battalion chief) were killed, and the other four were injured.

Marie Belle Scott-Uda fainted. An inquest was held, and the following statement was issued:

> We, the jury, find that the cause of the accident was in consequence of the ladder being made of inferior wood and its construction faulty; and we emphatically censure the Board of Fire Commissioners for not submitting the ladders to both a scientific and a practical test before allowing them to be used by the Fire Department. And we unhesitatingly condemn the further use of these aerial ladders by our Fire Department now owned by the City.

In due time, however, Hayes aerial ladders, rather than the Italian knock-offs, became an integral part of the New York Fire Department's arsenal of fire apparatus. As for William White, who made the unfortunate deal with Marie Belle Scott-Uda — he was fired.

THE FIREPROOF
THEATER DISASTER

*J*ust as the *Titanic* was advertised as unsinkable, so was Chicago's newest and plushiest theater, the Iroquois, advertised as fireproof. But, within a month after it had opened in 1903, the Iroquois Theater became the death pyre for 602 persons in the worst theater disaster in the history of the United States.

At about 3:30 p.m. on December 30th, the theater, authorized to hold 1,602 persons, was packed with 1,774 persons. Nearly 300 people were standing. The number of attendees included many children on Christmas vacation from school. They were watching the second act of "Mr. Blue Beard" when sparks from a carbon arc spotlight caught in the edge of a drapery near the light, high above the stage.

Comedian Eddie Foy, the star of the play, recalled, "When the blaze was first discovered, two stage hands tried to extinguish it. One of them strove to beat it out with a stick or a piece of canvas or something else, but it was too far above his head. Then he or the other man got one of those fire extinguishers consisting of a small tin tube of powder and tried to throw the stuff on the flames, but it was ridiculously inadequate."

Foy went on stage and shouted to the audience, "Don't get excited. There's no danger — take it easy." Then he yelled at the orchestra leader, "Play! Start an overture, anything, but *play!*"

The Iroquois had advertised an asbestos stage curtain. It was obviously time to use it, but as it started to come down it got stuck on a steel lighting reflector left in the way by a careless stagehand. One end of the curtain got within five feet of the stage, but the other end was stuck twenty feet in the air, which formed a flue for the deadly flames to sweep through.

Flammable stage props caught fire, and the smoke, trapped inside the theater, rose to the balcony, where hundreds began to gasp and choke. Of the 22 exits, the majority were not usable. According to Foy, "Few of them were marked by lights; some even had heavy portiers over the doors, and some of the doors were locked or fastened with levers which no one knew how to work."

The asbestos curtain, which never came down, was not asbestos at all. It was consumed by the fire and was found to actually have been made of burlap coated with asbestos paint.

About half of those who died perished near their seats, and the rest were either suffocated or trampled to death in the dark passages, stairways, and blind exits. Bodies were piled three and four deep.

The theater was described in the Mr. Blue Beard program as "absolutely fireproof." But within minutes of the sparks igniting the drapery, more than a third of the audience — 602 persons — were dead.

An article in *Fire and Water Engineering* stated, "Had it not been for the heroism of comedian Eddie Foy, who nearly lost his life and was severely burned in warning the audience while the flames surrounded him, the loss of life would have been still more terrible."

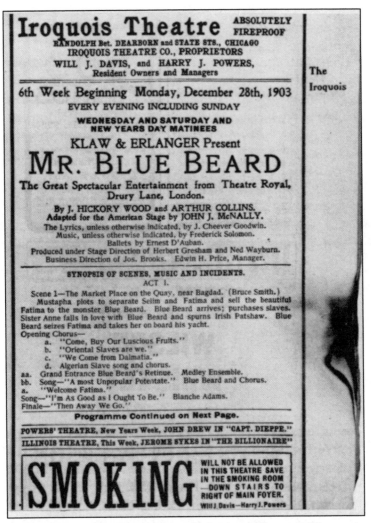

Within weeks after Chicago's new "absolutely fireproof" Iroquois Theater opened in 1903, more than 600 person lost their lives in America's most tragic theater fire. This scorched program was salvaged from the theater after the fire.

HOSE CARRIAGES
THAT NEVER
FOUGHT FIRES

In the "old days" of the early and mid 1800's, the various fire companies in cities were independent organizations, almost like social clubs, and the rivalry between them was usually intense.

Often fire companies would strive to own the most beautiful, ornate, and spectacular engines imaginable, replete with brass fittings, and ornamental lamps, while some even had side panels decorated with oil paintings by well known artists. Sometimes even these elaborate and fancy fire engines were not enough. Going to the extreme, some of the wealthiest fire companies commissioned the building of elaborate hose carriages, which were so magnificent that they were used only for "show." They remained inside the firehouses during fires, and were taken out only for parades. The following quotation describes some of Philadelphia's fanciest hose carriages:

> Every decoration which painters, sculptors, and labidaries could put upon them was used. They were resplendent with gold and silver work, handsome paintings, mirrored sides and carvings. They were

A side panel painting from the Americus Engine Co. 6 of New York City. Artist John Archibald Woodside painted the mythological scene, which he titled "Translation of Psyche."

inlaid with pearl, and one carriage bore on its front a blazing glory formed of imitation brilliants of the first order. The very handsome machines thus decorated seemed only designed for show, while the work was mostly done by uncouth, badly-shaped, clumsy carriages called "crabs," which bore as much resemblance to the dandy hose carriages as orang-outangs do to Venus.

Meanwhile, the New York volunteers were going all

This handsome and ornate hose carriage was never meant to carry hose nor to attend fires. It was built in 1884 for the Phoenix Hose Co. No. 1 of Poughkeepsie, New York, who used it only as a showpiece in parades. It sold at a Sotheby's Auction on August 18, 1993 for $79,500.

out with their lavish hose carriages also. Decorations included not only ornamental lamps but also side panels adorned with mirrors and velvet lining. Oil paintings included sea nymphs, goddesses, mermaids, and Indian princesses, some of them topless.

Perhaps the most expensive and elaborate of them all was the hose carriage of New York's Amity Hose Co. No. 38. It was made of the finest hard wood and painted white with blue stripes. It was adorned with heavy silver plating set off by two large trumpets and gold plated lamps with red lenses in the shape of pineapples. Paintings of mermaids and cupids were between all the glitter.

This hose carriage, paid for by wealthy New York businessmen who belonged to Amity Hose, cost eight thousand dollars — equal in today's money to nearly a quarter of a million dollars.

WHY ARE DALMATIONS
FIRE DOGS?

The answer is interesting, one you'll likely recall every time you see the Dalmatian/firehouse combo from now on.

According to the legend, it all began in the days of stagecoaches. Horse theft was so common back then that many stagecoach drivers strung a hammock be-

Why do Dalmations and firehouses go together like smoke and fire?

tween two stalls at night, then slept behind their horses to guard against thieves. But, if the driver owned a Dalmatian, he could sleep in the house or the stagecoach hotel. Why? Because it was observed that Dalmatians formed an amazingly tight bond with horses.

Since every firehouse had a set of fast horses to pull the pumper, it became common for each group of firemen to keep a Dalmatian. The spotted dogs not only guarded the firehouse horses, they kept them company during their long, boring waits between fires. And when they took off for a fire, the dog would run alongside the pumper.

The horses are gone from fire stations today, but the Dalmatians aren't. The tradition has been carried on, and it may be as much for the looks and appeal of these beautiful dogs as it is for their nostalgic tie to yesteryear.

Sticks, Stones,

And Fire Engines

John Wesley

The lives of brothers John and Charles Wesley, the original Methodists, were never easy. Born in England in 1703, John Wesley is credited not only with founding the Methodist Church and with preaching 40,560 sermons, but with having been set upon several times by angry mobs tryng to beat him, stone him, and kill him. Brother Charles fared little better.

But perhaps the most bizzare attempt on their lives occurred in 1747 at Devizes, England, when an angry mob attempted literally to flush Charles out of the house where he was preaching by using both of the village fire engines. John had a similar experience. But the Wesleys bested every mob that ever attacked them, including the ones who used fire engines.

The wrath of angry mobs was not an unusual experience during the 1700's for the Wesley brothers, John and Charles. Pursued by angry villagers in the English towns where they preached, the brothers often were pelted with sticks and stones.

But the weapons the mob used in Devizes, England, on February 25, 1747, were unlike any they ever had encountered. There the villagers came after Charles with both of the town's fire engines, determined to flush him out — first from the house where he was preaching, and next from the inn where he was staying. Here, taken from Charles Wesley's journal, is his own account of that eventful day:

"February 25th, (1747) - a day never to be forgotten. I walked quietly to Mrs. Phillips's and began preaching a little before the time appointed. Soon after, Satan's whole army assaulted the house. We sat in a little ground room, and ordered all the doors to be thrown open. They brought a hand engine, and began to play* into the house. We kept our seats, and they rushed into the passage; just then Mr. Borough, the constable, came, and seizing the spout of the engine, carried it off; but they hurried out to fetch the larger engine.

The rioters without continued to play their engine, which diverted them for some time; but their number and fierceness still increased; and the gentlemen supplied them with pitchers of ale, as much as they would drink. They were now on the point of breaking in, when Mr. Borough (the constable) though

---

*play: "To discharge, eject, or fire repeatedly or so as to make a stream, as hoses on a fire" This Webster definition of play is the one referenced by Charles Wesley, and is the derivation of playpipe, a fire service term used for over two centuries, although it is not listed in Webster's Unabridged Dictionary.

A hand fire engine of the type used in England in the mid-1700's. It was likely an engine such as this which Constable Borough "seized by the spout and carried it off." Note that the nozzle, or playpipe, was on a swivel or "gooseneck" and could be adjusted for placement of the stream without moving the engine.

of reading the proclamation. In less than the hour, of above a thousand wild beasts, none were left but the guard.

Our enemies at their return made their main assault at the back door, swearing horridly they would have me if it cost them their lives, many seeming accidents concurred to prevent their breaking in. They now got a notion that I had made my escape, and ran down to the inn and played the engine there. We were kept from all hurry and discomposure of spirit by a divine power resting upon us.

143

About three o'clock Mr. Clark knocked at the door, and brought with him the persecuting constable. He said, 'Sir, if you will promise never to preach here again, I will engage to bring you safe out of town.' My answer was, 'I shall promise no such thing.' The heart of our adversaries were turned.

While the constable was gathering his posse, we prepared to go forth. We rode a slow pace up the street, the whole multitude pouring along on both sides. Such fierceness and diabolical malice I have not before seen on human faces. When out of sight we mended our pace and joined in hearty prayer to our Deliverer."

Five years and one month later, on May 25, 1752, John Wesley, the more famous of the two brothers, made the following entry into *his* journal:

"We rode to Durham, and thence, through very rough roads, and as rough weather, to Barnard Castle. I was exceeding faint when we came in; however, the time being come, I went into the street, and would have preached; but the mob was so numerous and so loud that it was not possible for many to hear. Nevertheless I spoke on, and those who were near listened with huge attention. To prevent this, some of the rabble fetched the engine, and threw a good deal of water on the congregation; but not a drop fell on me."

Thirty-nine years later, at age 88, John Wesley preached his last sermon, number 40,560. The following is part of the inscription on a marble tablet erected to his memory in London:

"Regardless of fatigue, personal danger, and disgrace, he went out into the highways and hedges, calling sinners to repentance, and published the Gospel of Peace."

Starting with just the two Wesley brothers, John and Charles, the Methodist church had grown, at the time of John's death in 1791, to over 79,000 members, in spite of the mobs with sticks, stones, and fire engines. Today, the movement they founded more than 250 years ago has grown to number more than 50 million members around the world.

HONOLULU'S
BUBONIC PLAGUE FIRES

It is ordinarily unthinkable for firemen to become arsonists. Yet, in 1900, the president of the Honolulu Board of Health ordered the Honolulu Fire Department to set 41 separate fires in order to literally burn out the "Black Death" which was sweeping the city — the dreaded bubonic plague.

A Chinese bookkeeper, You Chong, working in the Asiatic Quarter of Honolulu, became ill on December 10, 1899, with a high fever and swelling in his groin. Called to Chong's bedside two days later, Dr. G.H. Herbert of the Board of Medical Examiners suspected the worst. The next day Chong was dead, and the autopsy confirmed Dr. Herbert's fear — bubonic plague.

That same day two more bubonic deaths were confirmed in the Asiatic Quarter. Schools were closed, rat guards were installed on all ships in port, cesspools were covered, streets were cleaned, rubbish was burned, and no ships were allowed to enter the harbor. Yet, on Christmas Eve, another case was reported, followed by

147

still another one on Christmas Day. Each following day additional cases were reported, and it became obvious that Honolulu was experiencing an epidemic of the Black Plague — the scourge that had, in the 14th century, wiped out one fourth of the entire population of Europe.

The Board of Health issued a proclamation: *If a building is in such an unsanitary condition that it cannot be purified by any means other than fire, then it should be destroyed by fire.*

On December 31, 1899, members of the Honolulu Fire Department set the first fire, which took down a row

Honolulu's Engine No. 1 was abandoned and destroyed as fire swept down Maunakea Street after the firemen themselves started the fire in an attempt to burn substandard housing thought to harbor infected rats. Photo courtesy of Hawaii State Archives.

of buildings on Nuuanu Avenue. The outbreak of plague continued. Next to be torched were the buildings surrounding the famous Kaumakapili Church where the late King David Kalakaua had worshipped. The fire chief assigned four steam fire engines to protect the church, but as the wind picked up and shifted, nearby buildings not scheduled for burning ignited, starting a conflagration the four engines could not stop. The church burned to the ground.

The fire burned out of control, consuming block after block of homes and businesses. A warehouse loaded with fireworks for the upcoming Chinese New Year festivities exploded. Buildings were dynamited in an attempt to form a fire break to halt the spread of the flames. Finally, seven hours after they had set the fire, Honolulu firefighters stopped it, but not before it had leveled 38 acres of buildings leaving 4,500 persons homeless.

Still the plague continued to spread, and before the Black Death was finally halted in April, firemen had set 31 additional fires. Although the plague itself claimed 61 lives, not a single life was lost because of the fires — fires set by the firemen themselves to burn out the bubonic plague.

EXTINGUISHERS OF YESTERDAY:

ALWAYS LETHAL,

NOW ILLEGAL

Less than two decades ago, millions of fire extinguishers in American factories, businesses and homes became not only obsolete, but illegal.

At least three out of every four fire extinguishers, including those carried on virtually all fire apparatus, used soda-acid, foam or carbon tetrachloride as the active ingredient. Although many of them were in use for nearly a century, all finally were outlawed as dangerous.

The soda-acid method of extinguishing fires generally is attributed to the French, but when in 1837 an American, Dr. William A. Graham, attempted to patent such an extinguisher, the U.S. Patent Office refused, advising that the device had "no practical use." Yet, half a century later, millions of soda-acid extinguishers hung in readiness in most industrial and commercial buildings throughout the United States.

These extinguishers usually held a two-and-one-half-gallon mixture of water and ordinary baking soda and were hung upright. Inside the copper or brass extinguisher shell was a glass bottle, with a loose stopper, filled with sulphuric acid or "oil of vitriol." Inverting the extinguisher caused the stopper to fall out and the acid to mix with the soda water. An immediate chemical reaction took place, creating carbon dioxide or "carbonic acid gas," which forced the soda water or "chemicals" out of a two-foot rubber hose with a one-eighth-inch tip.

Wild claims were made for soda-acid extinguishers, some true, but others false. Soda-acid extinguishers were: Dependable—true. Easy to use—true. Safe—false. They sometimes exploded, maiming or even killing the user. Thirty to 40 times more effective than plain water—false. The 1932 edition of the Fire Chief Handbook stated: "Contrary to general belief, the gas generated [in soda-acid extinguishers] adds little, if anything, to the extinguishing effect of the stream."

In 1972, soda-acid extinguishers finally were outlawed because of their propensity for exploding when used, although by that time, it was realized that stored pressure-water extinguishers were just as effective without the attendant mess of a soda water and sulphuric acid-charged stream .

Foam extinguishers, which appeared in the 1920s. were designed for flammable liquid fires as well as for the Class A fires controlled by their soda-acid cousins. They were similar to those containing soda acid, except that the acid was replaced with ferric or aluminum sulphate which, when mixed with the soda water, not only created carbon dioxide to force the mixture out of the nozzle, but created millions of tiny carbon dioxide

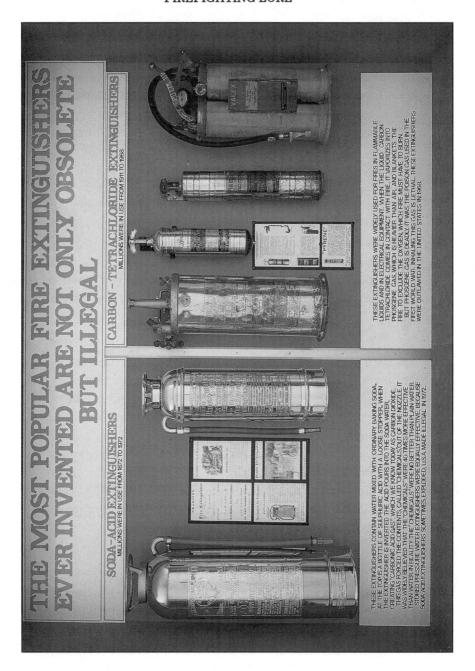

THE MOST POPULAR FIRE EXTINGUISHERS EVER INVENTED ARE NOT ONLY OBSOLETE BUT ILLEGAL

CARBON – TETRACHLORIDE EXTINGUISHERS
MILLIONS WERE IN USE FROM 1911 TO 1968

THESE EXTINGUISHERS WERE WIDELY USED FOR FIRES IN FLAMMABLE LIQUIDS AND IN ELECTRICAL EQUIPMENT. WHEN THE LIQUID CARBON TETRACHLORIDE COMES IN CONTACT WITH FIRE, IT VAPORIZES INTO PHOSGENE GAS, WHICH IS HEAVIER THAN AIR, AND BLANKETS THE FIRE TO EXCLUDE THE OXYGEN, WHICH FIRE MUST HAVE TO BURN. BUT PHOSGENE GAS IS DEADLY. IT WAS THE POISON GAS USED IN THE FIRST WORLD WAR. INHALING IT IS LETHAL. THESE EXTINGUISHERS WERE OUTLAWED IN THE UNITED STATES IN 1968.

SODA – ACID EXTINGUISHERS
MILLIONS WERE IN USE FROM 1872 TO 1972

THESE EXTINGUISHERS CONTAIN WATER MIXED WITH ORDINARY BAKING SODA. AT THE TOP IS A BOTTLE OF SULPHURIC ACID WITH A LOOSE STOPPER. WHEN THE EXTINGUISHER IS INVERTED, THE ACID POURS INTO THE SODA WATER, CREATING A CHEMICAL REACTION WHICH FORMS CARBON DIOXIDE. THIS GAS FORCED THE CONTENTS OUT. PEOPLE WHO KNEW SOME CHEMISTRY BELIEVED IT WAS WIDELY BELIEVED THAT THE "CHEMICALS" WERE 40 TIMES MORE EFFECTIVE THAN WATER. IN REALITY THE "CHEMICALS" WERE NO BETTER THAN PLAIN WATER. STORED PRESSURE WATER EXTINGUISHERS WERE EQUALLY EFFECTIVE. BECAUSE SODA-ACID EXTINGUISHERS SOMETIMES EXPLODED, U.S.A. MADE ILLEGAL IN 1972.

bubbles. These bubbles were toughened by adding a stabilizing agent to the soda solution. Again, they were phased out due to potential explosion.

The third kind of obsolete and outlawed fire extinguisher, the deadliest of all, was first made around 1910. The carbon tetrachloride extinguishers were intended for use on Class B and C fires (the liquid CCl_4, or "CTC" as it was often called, was a non-conductor of electricity). The CTC extinguisher did, indeed, smother fires with a heavier-than-air blanket of gas to exclude the oxygen. But inhaling the gas was lethal—it included the dreaded phosgene gas, which killed or impaired thousands of soldiers in World War I. It was, in effect, fighting fire with poison gas, since the phosgene would burn away minute protective hairs, or "cilia," within the user's lungs.

The "Pyrene" brand was the most popular of the carbon tetrachloride extinguishers, although there were dozens of competing manufacturers. Pyrene extinguishers took the form of small one-quart pump guns. There were various other sizes, including glass hand grenades. These grenades were thrown at the fire so they would break, releasing the carbon tetrachloride, which vaporized and smothered the fire. In 1968, little more than half a century after their introduction, the poisonous and sometimes lethal "carbon tet" extinguishers were outlawed in the United States.

THE STORY BEHIND

EMERGENCY

TELEPHONE STICKERS

The year was 1963, and the chief of the Perry Township Volunteer Fire Department near Evansville, Indiana, heard a siren approaching his fire station. As he glanced out the window, he saw, racing by on its way to a fire in his own township, Evansville Engine 7 with a puny 90 gallons of water in its booster tank. There were no hydrants in rural Perry Township, no alarm had been received, and the chief blew up. "This will never happen again," he promised himself. And it never did.

This was before the days of 9-1-1, and the telephone operators simply shunted all fire calls to the Evansville Fire Department. Soon the volunteer chief devised pressure-sensitive stickers made of brilliant red fluorescent label stock to fit in telephone cradles. On them was printed the number of the Perry Twp. Fire Department. "No one can overlook these," he thought. He thought right.

One of the original phone sticker advertisements.

The new emergency phone stickers were distributed to every home and business in Perry Township, and never again did the Perry firemen fail to receive an alarm. Neighboring volunteer fire departments heard about the stickers and ordered their own, with similar results. The Perry chief, an entrepreneur at heart, advertised them in state and, then, in national fire publications. Dozens of orders came in; but still not satisfied, the chief and his family stuck 26,000 actual red fluorescent phone stickers onto postcards bearing pictures of telephones and had them bound into a national fire department magazine.

The phone never stopped ringing for a month. The chief's wife, who handled the phone calls and paperwork, was overwhelmed. Before it was over, more than half of all the fire departments in the United States had ordered the new phone stickers. The printer turning out the stickers got hopelessly behind, and the Perry chief resigned his job of eleven years, installed a label printing press in the basement of his home, and printed them himself, often turning off the press to answer fire alarms.

Then requests for labels in various other shapes, sizes, and colors began to come in for a variety of uses. The label business boomed and outgrew the basement. Employees were hired, the operation expanded into a 5,000 square foot building, and still the growth continued.

Next, the chief's son graduated from college and opted to enter the label business with his father. He proved to be a mechanical genius, and invented label printing presses far superior to any that could be purchased. That took care of the extra demand for more labels, but the problem of handling all the paper work had also grown. To solve this problem the chief's daughter came home from college and assumed charge of administration. She, in turn, married a young man who proved to be a whiz at marketing. Within a decade, he had increased the company's dealer base from 500 to 30,000.

Today, the firm, known as Discount Labels®, has 400 employees, operates 60 label presses in an 160,000 square foot plant, and produces more orders of custom labels each day than any other label plant in the world.

Chief Conway, left, and his son Allen inspect several label orders. An average of 2,000 custom label orders are printed each day - more than any other label plant in the world.

This modern 160,000 square foot label plant has evolved from the first table-top printing press in Chief Conway's basement where emergency telephone stickers were first printed.

Three decades after devising the red fluorescent emergency telephone stickers, the fire chief turned label printer, decided to semi-retire in order to pursue his many avocations. Of these, his favorite is researching, writing, and publishing books about firefighting history. His name is W. Fred Conway — the author of this book.